D1573349

Radon Series on Computational and Applied Mathematics 6

Barbara Kaltenbacher
Andreas Neubauer
Otmar Scherzer

Iterative Regularization Methods for Nonlinear Ill-Posed Problems

Walter de Gruyter · Berlin · New York

Authors

Barbara Kaltenbacher
Institute of Stochastics and Applications
University of Stuttgart
Pfaffenwaldring 57
70569 Stuttgart
Germany
e-mail:
barbara.kaltenbacher@mathematik.uni-stuttgart.de

Andreas Neubauer
Industrial Mathematics Institute
Johannes Kepler University Linz
Altenbergerstraße 69
4040 Linz
Austria
e-mail: neubauer@indmath.uni-linz.ac.at

Otmar Scherzer
Institute of Mathematics
University of Innsbruck
ICT Building
Technikerstr. 21a
6020 Innsbruck
Austria

and

Johann Radon Institute for Computational
and Applied Mathematics (RICAM)
Austrian Academy of Sciences
Altenbergerstraße 69
4040 Linz
Austria
e-mail: otmar.scherzer@uibk.ac.at

Keywords

Nonlinear ill-posed problems, regularization, Landweber iteration, Newton type methods, multi-level methods, level set methods, discrepancy principle, parameter estimation problems, Radon transform.

Mathematics Subject Classification 2000

00-02, 47A52, 65-02, 65J15, 65J20, 65J22, 65N21.

♾ Printed on acid-free paper which falls within the guidelines
of the ANSI to ensure permanence and durability.

ISBN 978-3-11-020420-9

Bibliographic information published by the Deutsche Nationalbibliothek

The Deutsche Nationalbibliothek lists this publication in the Deutsche Nationalbibliografie; detailed bibliographic data are available in the Internet at http://dnb.d-nb.de.

Printed in Germany
Cover design: Martin Zech, Bremen.
Printing and binding: Hubert & Co. GmbH & Co. KG, Göttingen.

Preface

Driven by the needs of applications the field of inverse problems has been one of the fastest growing areas in applied mathematics in the last decades. It is well known that these problems typically lead to mathematical models that are *ill-posed*. This means especially that their solution is unstable under data perturbations.

Numerical methods that can cope with this problem are so-called *regularization methods*. The analysis of such methods for linear problems is relatively complete. The theory for nonlinear problems is developed to a much lesser extent. Several results on the well-known Tikhonov regularization and some results on Landweber iteration can be found in [45].

In the last years, more emphasis was put on the investigation of iterative regularization methods. It turned out that they are an attractive alternative to Tikhonov regularization especially for large scale nonlinear inverse problems. This book is devoted to the convergence and convergence rates analysis of iterative regularization methods like Landweber iteration and Newton type methods derived during the last ten years. In comparison to the recent monograph by Bakushinsky and Kokurin [5] we not only use standard smoothness conditions but also structural conditions on the equations under consideration and demonstrate, when they compensate for each other.

For aspects not covered by this book we refer the reader to other references. For instance we only give a deterministic but not a statistical error analysis; for the latter cf., e.g., [82, 155]. We also do not put our emphasis on specially designed computational methods for certain inverse problems such as parameter identification in PDEs of different types or inverse scattering. Here, we wish to refer the interested reader to corresponding references in [45] and the large number of recent publications on this subject.

We thank Herbert Egger (Linz), Markus Grasmair, Klaus Frick, Florian Frühauf, Richard Kowar, Frank Lenzen (Innsbruck), and Tom Lahmer (Erlangen) for their careful reading of parts of a preliminary draft of this book.

Each of us cooperated with several coauthors, who definitely influenced our understanding of the field. Special thanks are due to them.

We gratefully acknowledge financial support from the Austrian Fonds zur Förderung der wissenschaftlichen Forschung (projects Y-123INF, P15617, FSP S9203, S9207 and T7-TEC) and the German Forschungsgemeinschaft (project Ka 1778/1).

Barbara Kaltenbacher (Blaschke), Andreas Neubauer, and Otmar Scherzer

Preface

Contents

1 Introduction

Various feature articles concerned with *Inverse Problems* have been published recently in journals with high impact: in his article "In industry seeing is believing" West [159] describes the use of tomographical methods, such as X-ray, electromagnetic, and ultrasound tomography, for industrial applications like monitoring *oil pipelines.*

In their feature article "Mathematical methods in Imaging" Hero and Krim [70] discuss stochastic approaches for imaging. This article contains very early references from the 19th century to a problem of *image warping* which is a nonlinear inverse problem.

A feature story by Kowalenko [100] "Saving lives, one land mine at a time" describes recent research for land mine detection, where sensoring techniques like *ground penetration radar* and *electromagnetic induction* are used.

Another area of applications is medical imaging treated by Gould in "The rise and rise of medical imaging" [51] and by Vonderheid in "Seeing the invisible" [156].

It is a fact that many inverse problems are *ill-posed* in the sense that noise in measurement data may lead to significant misinterpretations of the solution. The ill-posedness can be handled either by incorporating a-priori information via the use of transformations, which stabilizes the problem, or by using appropriate numerical methods, called *regularization techniques.*

Typically, inverse problems are classified as *linear* or *nonlinear.* A classical example of a linear problem is *computerized tomography* (cf., e.g., Natterer [121]). The Radon transform, which is the basis of CT, has first been studied by Radon [133] in 1917 and until the first realization of a tomograph a huge amount of mathematical results have been developed. We refer to Webb [158] for a history of the development of CT scanners. Nowadays linear ill-posed problems still play an important role in inverse problems: we mention for instance the problem of *thermoacoustical tomography* (cf. [104, 105, 106, 160]).

Nonlinear inverse problems appear in a variety of natural models such as impedance tomography (cf., e.g., Borcea [12]) but also emerge, when for instance *geometrical* restrictions on the solution to be recovered are imposed: for highly unstable problems, it is advisable to constrain the degrees of freedom in the reconstruction according to physical principles, and for instance only recover the shapes of inclusions. Although the underlying problem may be linear (as for the inversion of the Radon transform) the geometrical constraints may make the problem nonlinear.

Due to rapidly evolving innovative processes in engineering and business, more and more new nonlinear inverse problems arise. However, in contrast to computerized tomography, a deep understanding of mathematical and physical aspects that would be necessary for deriving problem specific solution approaches can often not be gained for these new problems due to the lack of time. Therefore, one needs algorithms that can be used to solve inverse problems in their general formulation as nonlinear operator equations. It is the topic of this book to investigate such algorithms and provide a rigorous stability, convergence and convergence rates analysis.

1.1 Regularization methods

In this book, we treat problems given as nonlinear operator equations

$$F(x) = y\,, \tag{1.1}$$

where $F : \mathcal{D}(F) \to \mathcal{Y}$ with domain $\mathcal{D}(F) \subset \mathcal{X}$. We restrict our attention to Hilbert spaces $\mathcal{X}$ and $\mathcal{Y}$ with inner products $\langle \cdot , \cdot \rangle$ and norms $\| \cdot \|$, respectively; they can always be identified from the context in which they appear. Moreover, $\subset$ always denotes subset or equal, i.e., $A \subset A$.

Taking into account that in practice the data y are almost never available precisely, we denote the measured perturbed data by y^δ and assume that these *noisy* data satisfy

$$\|y^\delta - y\| \le \delta\,. \tag{1.2}$$

Problems of the form (1.1) that we have in mind are ill-posed in the sense that the solutions of (1.1) do not depend continuously on the data. Therefore, special methods, so-called *regularization methods*, are needed to get stable approximations of solutions of (1.1).

The probably most well-known method for solving nonlinear ill-posed problems is *Tikhonov regularization*: it consists in approximating a solution of (1.1) by a minimizer x_α^δ of the functional

$$x \mapsto \|F(x) - y^\delta\|^2 + \alpha\,\|x - x_0\|^2, \tag{1.3}$$

where $x_0 \in \mathcal{X}$ typically unifies all available a-priori information on the solution and α is a positive parameter.

Tikhonov regularization has been investigated extensively both for the solution of linear as well as nonlinear ill-posed problems (cf. [45] for a survey on continuous regularization methods and references therein). Under mild assumptions on the operator F it can be shown that, for $\alpha > 0$ fixed, the minimizers x_α^δ of the functional (1.3) are stable with respect to perturbations of the data y. Moreover, if (1.1) is solvable and if the regularization parameter $\alpha := \alpha(\delta)$ satisfies that $\alpha \to 0$ and that $\delta^2/\alpha \to 0$ as $\delta \to 0$, then x_α^δ converges to a solution of (1.1). In general, this convergence can be arbitrarily slow. Convergence rates results have been proven if F is Fréchet-differentiable and if the x_0-minimum-norm solution of (1.1) (denoted by $x^\dagger$) satisfies

$$x^\dagger - x_0 = (F'(x^\dagger)^* F'(x^\dagger))^\mu v$$

with $1/2 \le \mu \le 1$ and $\|v\|$ sufficiently small. If α is properly chosen in dependence on the *noise level* δ then the rate

$$\|x_\alpha^\delta - x^\dagger\| = O\left(\delta^{\frac{2\mu}{2\mu+1}}\right)$$

may be achieved.

1.2 Iterative regularization methods

The minimization of the Tikhonov functional for nonlinear ill-posed problems is usually realized via iterative methods. Since for linear ill-posed problems iterative regularization methods (see [45]) are an attractive alternative to Tikhonov regularization, we are interested in the regularization properties of iterative methods when applied to nonlinear problems.

In this book we concentrate on the numerical solution of (1.1) with iterative techniques of the form

$$x_{k+1}^\delta = x_k^\delta + G_k(x_k^\delta, y^\delta), \qquad k \in \mathbb{N},$$

for various choices of G_k. It turns out that under certain conditions the iteration scheme combined with an appropriate stopping criterion yields stable approximations of a solution of (1.1). The conditions for obtaining convergence and convergence rates results are more complicated than the ones needed in the analysis of Tikhonov regularization. This is due to the fact that the standard analysis of Tikhonov regularization does not incorporate a particular algorithm for finding the *global* minimizers. However, it is a-priori not guaranteed that the regularized solutions can be calculated with a convergent numerical algorithm that is not trapped in a local minimum. An analysis of convexity of the Tikhonov functional, guaranteeing global convergence of most numerical methods, has been performed by Chavent and Kunisch [27, 28, 29]. It has been pointed out in [66] that the conditions needed by Chavent and Kunisch and those necessary to prove convergence of iterative techniques are closely related.

For a linear problem $Kx = y$, where K is a linear operator between Hilbert spaces, many iterative methods for approximating $K^\dagger y$ (here $K^\dagger$ denotes the Moore–Penrose inverse, cf. Nashed [117]) are based on a transformation of the normal equation into an equivalent fixed point equation like, e.g.,

$$x = x + K^*(y - Kx)\,. \tag{1.4}$$

Since $K^*(y - Kx)$ is the direction of the negative gradient of the quadratic functional

$$\|y - Kx\|^2,$$

the appropriate fixed point equation for nonlinear problems is given by

$$x = \phi(x) := x + F'(x)^*(y - F(x)) \tag{1.5}$$

assuming that the nonlinear operator F is differentiable.

For well-posed problems convergence of iterative schemes is typically proven by fixed point arguments using *contraction* properties of the fixed point operator. This is true for instance for *descent algorithms* and *Newton type methods*. For ill-posed problems the situation is different, since there the operator ϕ is no contraction. The convergence theory for well-posed problems can be generalized to *nonexpansive* operators ϕ, i.e.,

$$\|\phi(x) - \phi(\tilde{x})\| \le \|x - \tilde{x}\|, \qquad x, \tilde{x} \in \mathcal{D}(\phi)\,.$$

Iterative methods for approximating fixed points of such operators have been considered in [3, 4, 16, 60, 130], to name just a few. In most of these references the major emphasis was put on a constructive proof of existence of fixed points of ϕ; Bakushinskii and Goncharskii [4] (see also [3, 5]) also considered the regularizing properties of such iterative schemes.

We believe that in many practical examples it is almost impossible to check analytically whether the operator ϕ is nonexpansive or not. Therefore, we replace the nonexpansivity of ϕ by local properties that are easier to verify and that guarantee at least local convergence of the iteration methods.

In the next two chapters we deal with the classical Landweber iteration and some modifications of it. In Chapter 4 we treat Newton type methods and in Chapter 5 multilevel methods. In Chapter 6 we discuss level set methods for inverse problems and how they can be realized via iterative regularization techniques. In Chapter 7 we present two numerical applications where we compare the perfomance of different iteration methods discussed in this book. Finally, in the last chapter we give comments on some other iterative regularization approaches that are not covered in this book.

2 Nonlinear Landweber iteration

As mentioned in Section 1.2, most iterative methods for linear ill-posed problems are based on the fixed point equation (1.4). In 1951, Landweber [110] proved strong convergence of the method of successive approximations applied to (1.4) for linear compact operators. An extensive study of this method including convergence rates results can be found, e.g., in [45, Section 6.1].

If we apply the method of successive approximations to the fixed point equation (1.5), we obtain a natural extension to nonlinear problems. Assuming throughout this chapter that F is a map between Hilbert spaces $\mathcal{X}$ and $\mathcal{Y}$, and that F has a continuous Fréchet-derivative $F'(\cdot)$, the *nonlinear Landweber iteration* is defined via

$$x_{k+1}^\delta = x_k^\delta + F'(x_k^\delta)^*(y^\delta - F(x_k^\delta)), \qquad k \in \mathbb{N}_0, \tag{2.1}$$

where y^δ are noisy data satisfying (1.2). By $x_0^\delta = x_0$ we denote an initial guess which may incorporate a-priori knowledge of an exact solution. If the Landweber iteration is applied to exact data, i.e., using y instead of y^δ in (2.1), then we write x_k instead of x_k^δ.

In case of noisy data, the iteration procedure has to be combined with a stopping rule in order to act as a regularization method. We will employ the *discrepancy principle*, i.e., the iteration is stopped after $k_* = k_*(\delta, y^\delta)$ steps with

$$\|y^\delta - F(x_{k_*}^\delta)\| \le \tau\delta < \|y^\delta - F(x_k^\delta)\|, \qquad 0 \le k < k_*, \tag{2.2}$$

where τ is an appropriately chosen positive number.

We mention that Morozov's discrepancy principle [116] – with $\tau > 1$ – has been applied successfully by Vainikko [152] to the regularization of linear ill-posed problems via Landweber iteration. In [34], Defrise and De Mol used a different technique to study the stopping rule (2.2) for $\tau > 2$.

The results on convergence and convergence rates presented in the next sections were established in [67].

2.1 Basic conditions

For nonlinear problems, iteration methods like (2.1) will in general not converge globally. We are able to prove local convergence if we impose some conditions on F.

As in the linear case, the Landweber iteration can only converge if problem (1.1) is properly scaled. For our analysis we assume that

$$\|F'(x)\| \leq 1, \qquad x \in \mathcal{B}_{2\rho}(x_0) \subset \mathcal{D}(F), \tag{2.3}$$

where $\mathcal{B}_{2\rho}(x_0)$ denotes a closed ball of radius 2ρ around x_0. Of course, instead of scaling the equation, in (2.1) one could also add a relaxation paramater to $F'(x_k^\delta)^*(y^\delta - F(x_k^\delta))$.

In addition to this scaling property, we need the following local condition:

$$\begin{aligned} &\|F(x) - F(\tilde{x}) - F'(x)(x - \tilde{x})\| \leq \eta \|F(x) - F(\tilde{x})\|, \qquad \eta < \tfrac{1}{2}, \\ &x, \tilde{x} \in \mathcal{B}_{2\rho}(x_0) \subset \mathcal{D}(F). \end{aligned} \tag{2.4}$$

Both conditions are strong enough to ensure local convergence to a solution of (1.1) if equation (1.1) is solvable in $\mathcal{B}_\rho(x_0)$. They also guarantee that all iterates x_k^δ, $0 \leq k \leq k_*$, remain in $\mathcal{D}(F)$, which makes the Landweber iteration well defined. Otherwise it would be necessary to project the iterates onto $\mathcal{D}(F)$ (cf., e.g., Vasin [154] and Eicke [43]).

From (2.4) it follows immediately with the triangle inequality that

$$\frac{1}{1+\eta}\|F'(x)(\tilde{x} - x)\| \leq \|F(\tilde{x}) - F(x)\| \leq \frac{1}{1-\eta}\|F'(x)(\tilde{x} - x)\| \tag{2.5}$$

for all $x, \tilde{x} \in \mathcal{B}_{2\rho}(x_0)$. Thus, condition (2.4) seems to be rather restrictive. However, the condition is quite natural as the following argument shows: if $F'(\cdot)$ is Lipschitz continuous and $x, \tilde{x} \in \mathcal{D}(F)$, then the error bound

$$\|F(\tilde{x}) - F(x) - F'(x)(\tilde{x} - x)\| \leq c\|\tilde{x} - x\|^2 \tag{2.6}$$

holds for the Taylor approximation of F. For ill-posed problems, however, it turns out that this estimate carries too little information about the local behaviour of F around x to draw conclusions about convergence of the nonlinear Landweber iteration, since the left hand side of (2.6) can be much smaller than the right hand side for certain pairs of points $\tilde{x}$ and x, whatever close to each other they are. For example, fix $x \in \mathcal{D}(F)$, and assume that F is weakly closed and compact. Then $F'(x)$ is compact, and hence, for every sequence $\{\tilde{x}_n\}$ with $\tilde{x}_n \in \mathcal{D}(F)$, $\|\tilde{x}_n - x\| = \varepsilon$ for all $n \in \mathbb{N}$, and $\tilde{x}_n \rightharpoonup x$ as $n \to \infty$, the left hand side of (2.6) goes to zero as $n \to \infty$ whereas the right hand side remains $c\varepsilon^2$ for all n.

For several examples one can even prove the stronger condition

$$\|F(x) - F(\tilde{x}) - F'(x)(x - \tilde{x})\| \leq c\|x - \tilde{x}\| \, \|F(x) - F(\tilde{x})\|. \tag{2.7}$$

Provided $\|x - \tilde{x}\|$ is sufficiently small, this implies condition (2.4).

To prove local convergence, we will always assume that the equation $F(x) = y$ is solvable in $\mathcal{B}_\rho(x_0)$. Note that then (2.5) even implies the existence of a unique solution of minimal distance to x_0. This so-called *x_0-minimum-norm solution* will be denoted by $x^\dagger$ (compare [45]). We show the following general result:

Proposition 2.1 *Let $\rho, \varepsilon > 0$ be such that*

$$\begin{aligned} &\|F(x) - F(\tilde{x}) - F'(x)(x - \tilde{x})\| \le c(x, \tilde{x}) \|F(x) - F(\tilde{x})\|, \\ &x, \tilde{x} \in \mathcal{B}_\rho(x_0) \subset \mathcal{D}(F), \end{aligned} \tag{2.8}$$

for some $c(x, \tilde{x}) \ge 0$, where $c(x, \tilde{x}) < 1$ if $\|x - \tilde{x}\| \le \varepsilon$.

(i) Then for all $x \in \mathcal{B}_\rho(x_0)$

$$M_x := \{\tilde{x} \in \mathcal{B}_\rho(x_0) \,:\, F(\tilde{x}) = F(x)\} = x + \mathcal{N}(F'(x)) \cap \mathcal{B}_\rho(x_0)$$

and $\mathcal{N}(F'(x)) = \mathcal{N}(F'(\tilde{x}))$ for all $\tilde{x} \in M_x$. Moreover,

$$\mathcal{N}(F'(x)) \supset \{t(\tilde{x} - x) \,:\, \tilde{x} \in M_x, t \in \mathbb{R}\},$$

where instead of $\supset$ equality holds if $x \in \mathring{\mathcal{B}}_\rho(x_0)$.

(ii) If $F(x) = y$ is solvable in $\mathcal{B}_\rho(x_0)$, then a unique x_0-minimum-norm solution exists. It is characterized as the solution $x^\dagger$ of $F(x) = y$ in $\mathcal{B}_\rho(x_0)$ satisfying the condition

$$x^\dagger - x_0 \in \mathcal{N}(F'(x^\dagger))^\perp. \tag{2.9}$$

Proof. Let $x, \tilde{x} \in \mathcal{B}_\rho(x_0)$ with $x \ne \tilde{x}$ and $F(x) = F(\tilde{x})$. Then (2.8) implies that $\tilde{x} - x \in \mathcal{N}(F'(x))$. If we set $x_t := x + t(\tilde{x} - x), t \in [0, 1]$, then $x_t - x \in \mathcal{N}(F'(x))$ and again by (2.8) we obtain that $F(x_t) = F(x)$ if $t \le \bar{t} := \varepsilon / \|\tilde{x} - x\|$. Moreover, $\tilde{x} - x \in \mathcal{N}(F'(x_t))$. If $\bar{t} < 1$, we have to repeat this procedure with x replaced by $x_{\bar{t}}$. After finitely many steps it then follows that

$$F(x_t) = F(x) = F(\tilde{x}) \quad \text{and} \quad \tilde{x} - x \in \mathcal{N}(F'(x_t)) \quad \text{for all} \quad t \in [0, 1]. \tag{2.10}$$

Let now $h \in \mathcal{N}(F'(x))$. Then there obviously exist $t \ne 0$ and $s \in \mathbb{R}$ such that $x_{t,s} := x + th + s(\tilde{x} - x)$ satisfies $x_{t,s} \in \mathcal{B}_\rho(x_0) \cap \mathcal{B}_\varepsilon(x)$. Now (2.8) and (2.10) imply that $F(x_{t,s}) = F(x) = F(\tilde{x})$ and that $x_{t,s} - \tilde{x} = th + (s - 1)(\tilde{x} - x) \in \mathcal{N}(F'(\tilde{x}))$. Thus, (2.10) yields that $h \in \mathcal{N}(F'(\tilde{x}))$. Changing roles of x and $\tilde{x}$ shows that

$$\mathcal{N}(F'(x)) = \mathcal{N}(F'(\tilde{x})). \tag{2.11}$$

On the other hand, if $h \in \mathcal{N}(F'(x))$ is such that $x_{t,0} \in \mathcal{B}_\rho(x_0)$ for some $t \ne 0$, then it follows as above that $F(x_{t,0}) = F(x)$ for all $t \in \mathbb{R}$ with $x_{t,0} \in \mathcal{B}_\rho(x_0)$.

This together with (2.10) and (2.11) implies all assertions of (i) noting that for $x \in \mathring{\mathcal{B}}_\rho(x_0)$ and $h \in \mathcal{N}(F'(x))$ there always exists a $t \neq 0$ such that $x_{t,0} \in \mathcal{B}_\rho(x_0)$.

We will now turn to the assertions in (ii) and assume that $F(x) = y$ is solvable in $\mathcal{B}_\rho(x_0)$. Then it follows with (i) that

$$M := \{x \in \mathcal{B}_\rho(x_0) \, : \, F(x) = y\}$$

is not empty, bounded, closed, and convex. Thus, M is weakly compact. Since the norm is weakly lower semicontinuous, an x_0-minimum-norm solution exists. By the convexity of M, the x_0-minimum-norm solution is unique and will be denoted by $x^\dagger$.

Let $h \in \mathcal{N}(F'(x^\dagger))$ and $x_t := x^\dagger + th$. Then it follows from above that $F(x_t) = F(x^\dagger)$ for all $t \in \mathbb{R}$ with $x_t \in \mathcal{B}_\rho(x_0)$. Therefore,

$$\|x_t - x_0\|^2 = \|x^\dagger - x_0\|^2 + t^2\|h\|^2 + 2t\langle x^\dagger - x_0, h \rangle > \|x^\dagger - x_0\|^2 \quad \text{if} \quad t \neq 0$$

and hence $\langle x^\dagger - x_0, h \rangle = 0$ if $t \neq 0$ exists with $x_t \in \mathcal{B}_\rho(x_0)$. If $x_t \notin \mathcal{B}_\rho(x_0)$ for all $t \neq 0$, then trivially $\langle x^\dagger - x_0, h \rangle = 0$. Thus, $x^\dagger - x_0 \in \mathcal{N}(F'(x^\dagger))^\perp$.

Let us now assume that $x^\dagger \in M$ satisfies (2.9) and that $x_* \in M$ with $x_* \neq x^\dagger$. Then (2.9) and (2.10) imply that

$$\begin{aligned} \|x_* - x_0\|^2 &= \|x_* - x^\dagger\|^2 + \|x^\dagger - x_0\|^2 + 2\langle x_* - x^\dagger, x^\dagger - x_0 \rangle \\ &= \|x_* - x^\dagger\|^2 + \|x^\dagger - x_0\|^2 > \|x^\dagger - x_0\|^2. \end{aligned}$$

Thus, $x^\dagger$ is the unique x_0-minimum-norm solution. □

If $F(x) = y$ is solvable in $\mathcal{B}_\rho(x_0)$ but a condition like (2.8) is not satisfied, then at least existence (but no uniqueness) of an x_0-minimum-norm solution is guaranteed provided that F is weakly sequentially closed (see [45, Chapter 10]).

2.2 Convergence of the Landweber iteration

Before we turn to a convergence analysis of the nonlinear Landweber iteration we want to emphasize that for fixed iteration index k the iterate x_k^δ depends continuously on the data y^δ, since x_k^δ is the result of a combination of continuous operations. This will be an important point in our analysis below; similar arguments were used for a general theory developed by Alifanov and Rumjancev [2].

To begin with, we formulate the following monotonicity property that gives us a clue how to choose the number τ in the stopping rule (2.2).

Proposition 2.2 *Assume that the conditions* (2.3) *and* (2.4) *hold and that the equation* $F(x) = y$ *has a solution* $x_* \in \mathcal{B}_\rho(x_0)$. *If* $x_k^\delta \in \mathcal{B}_\rho(x_*)$, *a sufficient condition for* x_{k+1}^δ *to be a better approximation of* x_* *than* x_k^δ *is that*

$$\|y^\delta - F(x_k^\delta)\| > 2\,\frac{1+\eta}{1-2\eta}\,\delta\,. \tag{2.12}$$

Moreover, it then holds that $x_k^\delta, x_{k+1}^\delta \in \mathcal{B}_\rho(x_*) \subset \mathcal{B}_{2\rho}(x_0)$.

Proof. Let us assume that $x_k^\delta \in \mathcal{B}_\rho(x_*)$. Then, due to the triangle inequality, $x_*, x_k^\delta \in \mathcal{B}_{2\rho}(x_0)$. Hence, (2.3) and (2.4) are applicable and we obtain together with definition (2.1) and (1.2) that

$$\begin{aligned}
&\|x_{k+1}^\delta - x_*\|^2 - \|x_k^\delta - x_*\|^2 \\
&\quad= 2\langle\, x_{k+1}^\delta - x_k^\delta, x_k^\delta - x_*\,\rangle + \|x_{k+1}^\delta - x_k^\delta\|^2 \\
&\quad= 2\langle\, y^\delta - F(x_k^\delta), F'(x_k^\delta)(x_k^\delta - x_*)\,\rangle + \|F'(x_k^\delta)^*(y^\delta - F(x_k^\delta))\|^2 \\
&\quad\le 2\langle\, y^\delta - F(x_k^\delta), y^\delta - F(x_k^\delta) - F'(x_k^\delta)(x_* - x_k^\delta)\,\rangle - \|y^\delta - F(x_k^\delta)\|^2 \\
&\quad\le \|y^\delta - F(x_k^\delta)\|\,(2\delta + 2\eta\,\|y - F(x_k^\delta)\| - \|y^\delta - F(x_k^\delta)\|\,) \\
&\quad\le \|y^\delta - F(x_k^\delta)\|\,(2(1+\eta)\delta - (1-2\eta)\,\|y^\delta - F(x_k^\delta)\|\,)\,.
\end{aligned} \tag{2.13}$$

The assertions now follow since the right hand side is negative if (2.12) holds. □

In view of this proposition, the number τ in the stopping rule (2.2) should be chosen subject to the following constraint depending on η, with η as in (2.4):

$$\tau > 2\,\frac{1+\eta}{1-2\eta} > 2\,. \tag{2.14}$$

From the proof of Proposition 2.2 we can easily extract an inequality that guarantees that the stopping index k_* in (2.2) is finite and hence well defined.

Corollary 2.3 *Let the assumptions of Proposition* 2.2 *hold and let* k_* *be chosen according to the stopping rule* (2.2), (2.14). *Then*

$$k_*(\tau\delta)^2 < \sum_{k=0}^{k_*-1} \|y^\delta - F(x_k^\delta)\|^2 \le \frac{\tau}{(1-2\eta)\tau - 2(1+\eta)}\,\|x_0 - x_*\|^2. \tag{2.15}$$

In particular, if $y^\delta = y$ *(i.e., if* $\delta = 0$*), then*

$$\sum_{k=0}^{\infty} \|y - F(x_k)\|^2 < \infty\,. \tag{2.16}$$

Proof. Since $x_0^\delta = x_0 \in \mathcal{B}_\rho(x_*)$, it follows by induction that Proposition 2.2 is applicable for all $0 \le k < k_*$. From (2.13) we then conclude that

$$\|x_{k+1}^\delta - x_*\|^2 - \|x_k^\delta - x_*\|^2 \le \|y^\delta - F(x_k^\delta)\|^2(2\tau^{-1}(1+\eta) + 2\eta - 1)\,.$$

Adding up these inequalities for k from 0 through $k_* - 1$, we obtain

$$(1 - 2\eta - 2\tau^{-1}(1+\eta)) \sum_{k=0}^{k_*-1} \|y^\delta - F(x_k^\delta)\|^2 \le \|x_0 - x_*\|^2 - \|x_{k_*}^\delta - x_*\|^2$$

which together with (2.2) yields (2.15). Obviously, if $\delta = 0$ then k_* may be any positive integer in (2.15) and τ may be chosen arbitrarily large yielding the estimate

$$\sum_{k=0}^{\infty} \|y - F(x_k)\|^2 \le \frac{1}{(1-2\eta)} \|x_0 - x_*\|^2.$$

□

Note that (2.16) implies that, if Landweber iteration is run with precise data $y = y^\delta$, then the residual norms of the iterates tend to zero as $k \to \infty$. That is, if the iteration converges, then the limit is necessarily a solution of $F(x) = y$. We will prove this convergence in the next theorem.

Theorem 2.4 *Assume that the conditions* (2.3) *and* (2.4) *hold and that* $F(x) = y$ *is solvable in* $\mathcal{B}_\rho(x_0)$. *Then the nonlinear Landweber iteration applied to exact data* y *converges to a solution of* $F(x) = y$. *If* $\mathcal{N}(F'(x^\dagger)) \subset \mathcal{N}(F'(x))$ *for all* $x \in \mathcal{B}_\rho(x^\dagger)$, *then* x_k *converges to* $x^\dagger$ *as* $k \to \infty$.

Proof. We know from Section 2.1 that the unique x_0-minimum-norm solution, $x^\dagger$, exists in $\mathcal{B}_\rho(x_0)$. Let

$$e_k := x_k - x^\dagger\,,$$

then Proposition 2.2 implies that $\|e_k\|$ monotonically decreases to some $\varepsilon \ge 0$. We are going to show that $\{e_k\}$ is a Cauchy sequence. Given $j \ge k$ we choose some integer l between k and j with

$$\|y - F(x_l)\| \le \|y - F(x_i)\| \qquad \text{for all} \qquad k \le i \le j\,. \tag{2.17}$$

We have

$$\|e_j - e_k\| \le \|e_j - e_l\| + \|e_l - e_k\| \tag{2.18}$$

and

$$\begin{aligned} \|e_j - e_l\|^2 &= 2\langle e_l - e_j, e_l\rangle + \|e_j\|^2 - \|e_l\|^2, \\ \|e_l - e_k\|^2 &= 2\langle e_l - e_k, e_l\rangle + \|e_k\|^2 - \|e_l\|^2. \end{aligned} \tag{2.19}$$

For $k \to \infty$, the last two terms on each of the right hand sides of (2.19) converge to $\varepsilon^2 - \varepsilon^2 = 0$. We now apply (2.1) and (2.4) to show that $\langle e_l - e_j, e_l \rangle$ also tends to zero as $k \to \infty$:

$$\begin{aligned}
|\langle e_l - e_j, e_l \rangle| &= \Big| \sum_{i=l}^{j-1} \langle F'(x_i)^*(y - F(x_i)), e_l \rangle \Big| \\
&\leq \sum_{i=l}^{j-1} |\langle y - F(x_i), F'(x_i)(x_l - x^\dagger) \rangle| \\
&\leq \sum_{i=l}^{j-1} \|y - F(x_i)\| \, \|F'(x_i)(x_l - x_i + x_i - x^\dagger)\| \\
&\leq \sum_{i=l}^{j-1} \|y - F(x_i)\| \Big(\|y - F(x_i) - F'(x_i)(x^\dagger - x_i)\| \\
&\qquad + \|y - F(x_l)\| + \|F(x_i) - F(x_l) - F'(x_i)(x_i - x_l)\| \Big) \\
&\leq (1+\eta) \sum_{i=l}^{j-1} \|y - F(x_i)\| \, \|y - F(x_l)\| + 2\eta \sum_{i=l}^{j-1} \|y - F(x_i)\|^2 \\
&\leq (1+3\eta) \sum_{i=l}^{j-1} \|y - F(x_i)\|^2,
\end{aligned}$$

where we have used (2.17) to obtain the last inequality. Similarly, one can show that

$$|\langle e_l - e_k, e_l \rangle| \leq (1+3\eta) \sum_{i=k}^{l-1} \|y - F(x_i)\|^2 .$$

With these estimates it follows from (2.16) that the right hand sides of (2.19) go to zero as $k \to \infty$. Thus, it follows with (2.18) that $\{e_k\}$ and hence also $\{x_k\}$ are Cauchy sequences. Finally, in view of the remark following Corollary 2.3, the limit of x_k as $k \to \infty$ must be a solution of $F(x) = y$.

If $\mathcal{N}(F'(x^\dagger)) \subset \mathcal{N}(F'(x))$ for all $x \in \mathcal{B}_\rho(x^\dagger)$, then by the definition (2.1) of the iterates

$$x_{k+1} - x_k \in \mathcal{R}(F'(x_k)^*) \subset \mathcal{N}(F'(x_k))^\perp \subset \mathcal{N}(F'(x^\dagger))^\perp$$

and hence

$$x_k - x_0 \in \mathcal{N}(F'(x^\dagger))^\perp \quad \text{for all} \quad k \in \mathbb{N} .$$

Therefore, this then also holds for the limit of x_k. Since $x^\dagger$ is the unique solution for which this condition holds (cf. (2.9)), this proves that $x_k \to x^\dagger$ as $k \to \infty$. □

We mention that the proof of x_k to be a Cauchy sequence is similar to an argument by McCormick and Rodrigue [114], that they have applied to prove convergence of the method of steepest descent for *linear problems*.

Remark 2.5 We emphasize that, in general, the limit of the Landweber iterates is no x_0-minimum-norm solution. However, since the monotonicity result of Proposition 2.2 holds for every solution, the limit of x_k has to be at least close to $x^\dagger$. As can be seen below, it has to be the closer the larger ρ can be chosen.

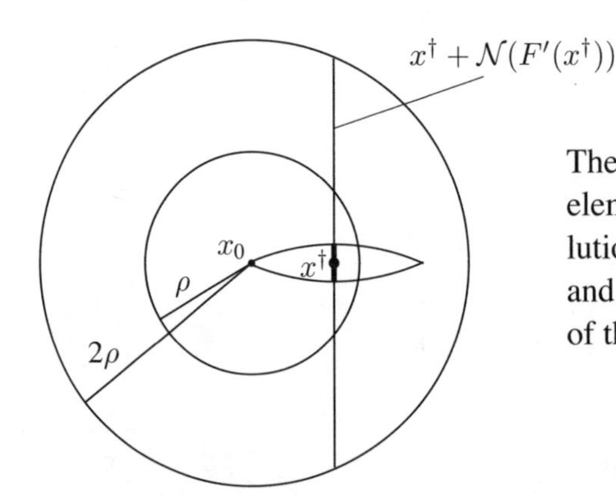

The sketch on the left shows the initial element x_0, the x_0-minimum-norm solution $x^\dagger$, the subset $x^\dagger + \mathcal{N}(F'(x^\dagger))$ and in bold the region, where the limit of the iterates x_k can be.

It is well known that, if y^δ does not belong to the range of F, then the iterates x_k^δ of (2.1) cannot converge but still allow a stable approximation of a solution of $F(x) = y$ provided the iteration is stopped after k_* steps. The next result shows that the stopping rule (2.2), (2.14) renders the Landweber iteration a regularization method.

Theorem 2.6 *Let the assumptions of Theorem* 2.4 *hold and let* $k_* = k_*(\delta, y^\delta)$ *be chosen according to the stopping rule* (2.2), (2.14). *Then the Landweber iterates* $x_{k_*}^\delta$ *converge to a solution of* $F(x) = y$. *If* $\mathcal{N}(F'(x^\dagger)) \subset \mathcal{N}(F'(x))$ *for all* $x \in \mathcal{B}_\rho(x^\dagger)$, *then* $x_{k_*}^\delta$ *converges to* $x^\dagger$ *as* $\delta \to 0$.

Proof. Let x_* be the limit of the Landweber iteration with precise data y and let $\{\delta_n\}$ be a sequence converging to zero as $n \to \infty$. Denote by $y_n := y^{\delta_n}$ a corresponding sequence of perturbed data, and by $k_n = k_*(\delta_n, y_n)$ the stopping index determined from the discrepancy principle for the Landweber iteration applied to the pair (δ_n, y_n).

Assume first that k is a finite accumulation point of $\{k_n\}$. Without loss of generality, we can assume that $k_n = k$ for all $n \in \mathbb{N}$. Thus, from the definition of k_n it follows that

$$\|y_n - F(x_k^{\delta_n})\| \le \tau\delta_n\,. \tag{2.20}$$

As k is fixed, x_k^δ depends continuously on y^δ, and hence, if we go to the limit $n \to \infty$ in (2.20), we obtain

$$x_k^{\delta_n} \to x_k\,, \quad F(x_k^{\delta_n}) \to F(x_k) = y \quad \text{as} \quad n \to \infty\,.$$

In other words, the kth iterate of the Landweber iteration with precise data is a solution of $F(x) = y$, and hence, the iteration terminates with $x_* = x_k$, and $x_{k_n}^{\delta_n} \to x_*$ for this subsequence $\delta_n \to 0$.

It remains to consider the case where $k_n \to \infty$ as $n \to \infty$. Then, for $k_n > k$ Proposition 2.2 yields

$$\|x_{k_n}^{\delta_n} - x_*\| \le \|x_k^{\delta_n} - x_*\| \le \|x_k^{\delta_n} - x_k\| + \|x_k - x_*\|\,. \tag{2.21}$$

Given $\varepsilon > 0$ it follows from Theorem 2.4 that we can fix some $k = k(\varepsilon)$ so large that the second term on the right hand side of (2.21) is smaller than $\varepsilon/2$. Because of the stability of the nonlinear Landweber iteration we also have that $\|x_k^{\delta_n} - x_k\| < \varepsilon/2$ for all $n > n(\varepsilon, k)$, showing that the left hand side of (2.21) is smaller than ε for n sufficiently large (so that $k_n > k$). Thus, $x_{k_n}^{\delta_n} \to x_*$ as $n \to \infty$. □

2.3 Convergence rates

It is well known that, under the general assumptions of the previous section, the rate of convergence of $x_k \to x_*$ as $k \to \infty$ (with precise data) or $x_{k_*}^\delta \to x_*$ as $\delta \to 0$ (with perturbed data) will, in general, be arbitrarily slow. For linear ill-posed problems $Kx = y$ convergence rates are obtained if the following *source conditions* are satisfied (cf. [45]):

$$x^\dagger - x_0 = (K^*K)^\mu v\,, \qquad \mu > 0\,, \quad v \in \mathcal{N}(K)^\perp.$$

For nonlinear problems, the corresponding condition is given by (cf. [45])

$$x^\dagger - x_0 = (F'(x^\dagger)^* F'(x^\dagger))^\mu v\,, \quad v \in \mathcal{N}(F'(x^\dagger))^\perp. \tag{2.22}$$

In many examples, this condition implies a certain *smoothness* of $x^\dagger - x_0$ if $\mu > 0$.

In contrast to Tikhonov regularization, assumption (2.22) (with $\|v\|$ sufficiently small) is not enough to obtain convergence rates for Landweber iteration. In [67] rates were proven under the additional assumption that F satisfies

$$F'(x) = R_x F'(x^\dagger) \quad \text{and} \quad \|R_x - I\| \le c\|x - x^\dagger\|\,, \qquad x \in \mathcal{B}_{2\rho}(x_0)\,, \tag{2.23}$$

where $\{R_x : x \in \mathcal{B}_{2\rho}(x_0)\}$ is a family of bounded linear operators $R_x : \mathcal{Y} \to \mathcal{Y}$ and c is a positive constant.

Unfortunately, the conditions above are not always satisfied (see [67, Example 4.3]). To enlarge the applicability of the results, in this section we consider instead of (2.1) the following slightly modified iteration method,

$$x_{k+1}^\delta = x_k^\delta + \omega G^\delta(x_k^\delta)^*(y^\delta - F(x_k^\delta)), \quad k \in \mathbb{N}_0, \tag{2.24}$$

where, as above, $x_0^\delta = x_0$ is an initial guess, $G^\delta(x) := G(x, y^\delta)$, and G is a continuous operator mapping $\mathcal{D}(F) \times \mathcal{Y}$ into $\mathcal{L}(\mathcal{X}, \mathcal{Y})$. The iteration will again be stopped according to the discrepancy principle (2.2).

Another modification were the operator F is approximated by a sequence of operators F_k that are possibly easier to evaluate was considered in [134].

To obatin local convergence and convergence rates for the modification (2.24) we need the following assumptions:

Assumption 2.7 Let ρ be a positive number such that $\mathcal{B}_{2\rho}(x_0) \subset \mathcal{D}(F)$.

(i) The equation $F(x) = y$ has an x_0-minimum-norm solution $x^\dagger$ in $\mathcal{B}_\rho(x_0)$.

(ii) There exist positive constants c_1, c_2, c_3 and linear operators R_x^δ such that for all $x \in \mathcal{B}_\rho(x^\dagger)$ the following estimates hold:

$$\|F(x) - F(x^\dagger) - F'(x^\dagger)(x - x^\dagger)\| \le c_1 \|F(x) - F(x^\dagger)\| \, \|x - x^\dagger\|, \tag{2.25}$$

$$G^\delta(x) = R_x^\delta G^\delta(x^\dagger), \tag{2.26}$$

$$\|R_x^\delta - I\| \le c_2 \|x - x^\dagger\|, \tag{2.27}$$

$$\|F'(x^\dagger) - G^\delta(x^\dagger)\| \le c_3 \delta. \tag{2.28}$$

(iii) The scaling parameter ω in (2.24) satisfies the condition

$$\omega \|F'(x^\dagger)\|^2 \le 1. \tag{2.29}$$

Note that, if instead of (2.25) the slightly stronger condition (2.7) holds in $\mathcal{B}_{2\rho}(x_0)$, then the unique existence of the x_0-minimum-norm solution $x^\dagger$ follows from Proposition 2.1 if $F(x) = y$ is solvable in $\mathcal{B}_\rho(x_0)$.

The next theorem states that under Assumption 2.7 the modified Landweber iteration (2.24) converges locally to $x^\dagger$ if τ is chosen properly:

Theorem 2.8 *Let Assumption* 2.7 *hold and let $k_* = k_*(\delta, y^\delta)$ be chosen according to the stopping rule* (2.2). *Moreover, we assume that $\|x_0 - x^\dagger\|$ is so small and that the parameter τ in* (2.2) *is so large that*

$$2\eta_1 + \eta_2^2\eta_3^2 < 2 \tag{2.30}$$

and

$$\tau > \frac{2(1 + \eta_1 + c_3\eta_2\,\|x_0 - x^\dagger\|)}{2 - 2\eta_1 - \eta_2^2\eta_3^2}\,, \tag{2.31}$$

where

$$\begin{aligned}
\eta_1 &:= \|x_0 - x^\dagger\|\,(c_1 + c_2(1 + c_1\,\|x_0 - x^\dagger\|))\,,\\
\eta_2 &:= 1 + c_2\,\|x_0 - x^\dagger\|\,,\\
\eta_3 &:= 1 + 2c_3\,\|x_0 - x^\dagger\|\,.
\end{aligned}$$

Then the modified Landweber iterates $x_{k_*}^\delta$ converge to $x^\dagger$ as $\delta \to 0$.

Proof. The proof follows essentially the lines of the last section: the results of Proposition 2.2, Corollary 2.3, Theorem 2.4 and Theorem 2.6 are also valid for the modified Landweber iteration. We only show the major differences of the proofs.

To simplify the notation we set $K := F'(x^\dagger)$ and $e_k^\delta := x_k^\delta - x^\dagger$. We show in a first step that $\|e_k^\delta\|$ is monotonically decreasing for $k \le k_*$.

Let us assume that $x_k^\delta \in \mathcal{B}_\rho(x^\dagger)$. Then it follows with (2.24) and (1.2) that

$$\begin{aligned}
\|e_{k+1}^\delta\|^2 - \|e_k^\delta\|^2 &= 2\omega\langle\, y^\delta - F(x_k^\delta), G^\delta(x_k^\delta)e_k^\delta\,\rangle\\
&\quad + \omega^2\,\|G^\delta(x_k^\delta)^*(y^\delta - F(x_k^\delta))\|^2\\
&\le \omega\,\|y^\delta - F(x_k^\delta)\|\,(2\delta + 2\,\|F(x^\dagger) - F(x_k^\delta) + G^\delta(x_k^\delta)e_k^\delta\|\\
&\quad - 2\,\|y^\delta - F(x_k^\delta)\| + \omega\,\|G^\delta(x_k^\delta)\|^2\,\|y^\delta - F(x_k^\delta)\|)\,.
\end{aligned}$$

The conditions (2.25) – (2.28) yield the estimates

$$\begin{aligned}
&\|F(x^\dagger) - F(x_k^\delta) + G^\delta(x_k^\delta)e_k^\delta\|\\
&\quad\le \|F(x_k^\delta) - F(x^\dagger) - Ke_k^\delta\| + \|(I - R_{x_k^\delta}^\delta)Ke_k^\delta\|\\
&\quad\quad + \|R_{x_k^\delta}^\delta(K - G^\delta(x^\dagger))e_k^\delta\|\\
&\quad\le \|e_k^\delta\|\,\Big(\,\|y - F(x_k^\delta)\|\,(c_1 + c_2(1 + c_1\,\|e_k^\delta\|)) + c_3\delta(1 + c_2\,\|e_k^\delta\|)\Big)
\end{aligned}$$

and

$$\|G^\delta(x_k^\delta)\| \le (1 + c_2\,\|e_k^\delta\|)(\|K\| + c_3\delta)\,.$$

Since (2.30) and (2.31) imply that $\tau > 2$, it follows with (1.2), (2.2), and (2.25) that for $k < k_*$ and $c_1\,\|e_k^\delta\| < 1$

$$\delta \le \frac{\|Ke_k^\delta\|}{1 - c_1\,\|e_k^\delta\|}\,.$$

Combining the estimates above with (1.2) and (2.29) we obtain

$$\begin{aligned}\|e_{k+1}^\delta\|^2 - \|e_k^\delta\|^2 \leq\; & \omega\|y^\delta - F(x_k^\delta)\|\Big(2\delta(1+\eta_{k,1}^\delta + c_3\eta_{k,2}^\delta\|e_k^\delta\|) \\ & - \|y^\delta - F(x_k^\delta)\|(2 - 2\eta_{k,1}^\delta - (\eta_{k,2}^\delta\eta_{k,3}^\delta)^2)\Big),\end{aligned} \tag{2.32}$$

where

$$\begin{aligned}\eta_{k,1}^\delta &:= \|e_k^\delta\|(c_1 + c_2(1 + c_1\|e_k^\delta\|))\,,\\ \eta_{k,2}^\delta &:= 1 + c_2\|e_k^\delta\|\,,\\ \eta_{k,3}^\delta &:= 1 + \frac{c_3\|e_k^\delta\|}{1 - c_1\|e_k^\delta\|}\,.\end{aligned}$$

Since $x_0^\delta = x_0 \in \mathcal{B}_\rho(x^\dagger)$ and since, by virtue of (2.30), $c_1\|e_0\| \leq 1/2$, it now follows together with (2.32), (2.2), and (2.31) that $\|e_1^\delta\| < \|e_0^\delta\| = \|e_0\|$, as long as $k_* > 0$, and that estimate (2.32) is applicable with $k = 1$ if $1 < k_*$.

Let us now assume that $\|e_0\| > \|e_1^\delta\| > \ldots > \|e_k^\delta\|$ for some $0 \leq k < k_*$. Then, by monotonicity, it follows from (2.32) that

$$\begin{aligned}\|e_{k+1}^\delta\|^2 - \|e_k^\delta\|^2 \leq\; & \omega\|y^\delta - F(x_k^\delta)\|\Big(2\delta(1+\eta_1 + c_3\eta_2\|e_0\|) \\ & - \|y^\delta - F(x_k^\delta)\|(2 - 2\eta_1 - \eta_2^2\eta_3^2)\Big)\end{aligned}$$

implying as above that $\|e_{k+1}^\delta\| < \|e_k^\delta\|$. Thus, we have shown by induction that $\|e_k^\delta\|$ is monotonically decreasing for $k \leq k_*$.

Moreover, it follows that a similar estimate like (2.15) holds that guarantees that the stopping index k_* is finite if $\delta > 0$ and that

$$\sum_{k=0}^{\infty}\|y - F(x_k)\|^2 < \infty\,.$$

Now we show that also the results of Theorem 2.4 are valid. Note that due to (2.25) – (2.28) we get the estimate

$$\|G^0(x_i)(x_l - x^\dagger)\| \leq (1 + c_2\|x_i - x^\dagger\|)(1 + c_1\|x_l - x^\dagger\|)\,\|y - F(x_l)\|\,.$$

Together with (2.17) and the monotonicity of $\|x_i - x^\dagger\|$ we then obtain that

$$\begin{aligned}|\langle e_l - e_j, e_l\rangle| &= \omega\Big|\sum_{i=l}^{j-1}\langle G^0(x_i)^*(y - F(x_i)), x_l - x^\dagger\rangle\Big| \\ &\leq \omega(1 + c_2\|x_0 - x^\dagger\|)(1 + c_1\|x_0 - x^\dagger\|)\sum_{i=l}^{j-1}\|y - F(x_i)\|^2.\end{aligned}$$

This yields convergence of x_k towards a solution $x \in \mathcal{B}_\rho(x^\dagger)$ and, due to (2.26) and (2.28), that $x - x_0 \in \mathcal{N}(F'(x^\dagger))^\perp$. Since (2.25) implies that $x - x^\dagger \in \mathcal{N}(F'(x^\dagger))$, we obtain that

$$\|x^\dagger - x_0\|^2 = \|x^\dagger - x\|^2 + \|x - x_0\|^2.$$

Since $x^\dagger$ is an x_0-minimum-norm solution, this implies that $x = x^\dagger$.

The convergence of $x_{k_*}^\delta$ towards $x^\dagger$ as $\delta \to 0$ follows as in the proof of Theorem 2.6 noting that due to the continuity of G the modified Landweber iterates x_k^δ again depend continuously on y^δ. □

If we, in addition to Assumption 2.7, require that $x^\dagger$ satisfies the source condition (2.22), then we even obtain convergence rates. For the proof we need the following lemmata:

Lemma 2.9 *Let p and q be nonnegative. Then there is a positive constant $c(p,q)$ independent of k so that*

$$\sum_{j=0}^{k-1} (j+1)^{-p}(k-j)^{-q} \le c(p,q)(k+1)^{1-p-q}h(k)$$

with

$$h(k) := \begin{cases} 1\,, & \max\{p,q\} < 1\,, \\ \ln(k+1)\,, & \max\{p,q\} = 1\,, \\ (k+1)^{\max\{p,q\}-1}\,, & \max\{p,q\} > 1\,. \end{cases}$$

Proof. Let us first assume that $q = 0$ (the case $p = 0$ is similar). Then the following estimate

$$\sum_{j=0}^{k-1} (j+1)^{-p} \le 1 + \int_1^k x^{-p}\,dx = \begin{cases} 1 + (1-p)^{-1}(k^{1-p} - 1)\,, & p \ne 1\,, \\ 1 + \ln(k)\,, & p = 1\,, \end{cases}$$

holds, which immediately yields the assertion.

Now assume that $p > 0$ and $q > 0$. Since the function $f(s) := s^{-p}(1-s)^{-q}$ is convex in $(0,1)$, we obtain that

$$\begin{aligned} &\sum_{j=0}^{k-1} (j+1)^{-p}(k-j)^{-q} \\ &\quad = (k+1)^{1-p-q} \sum_{j=0}^{k-1} \left(\frac{j+1}{k+1}\right)^{-p} \left(1 - \frac{j+1}{k+1}\right)^{-q} \frac{1}{k+1} \\ &\quad \le (k+1)^{1-p-q} \int_h^{1-h} s^{-p}(1-s)^{-q}\,ds \end{aligned}$$

$$\leq \ (k+1)^{1-p-q}\left(2^q\int_h^{\frac{1}{2}} s^{-p}\,ds + 2^p\int_{\frac{1}{2}}^{1-h}(1-s)^{-q}\,ds\right)$$

for $h := 1/(2k+2)$. Now the assertion follows from the last estimate. □

Lemma 2.10 *Let $K \in \mathcal{L}(\mathcal{X},\mathcal{Y})$ and $\omega > 0$ be such that $\omega\|K\|^2 \leq 1$ and let $s \in [0,1]$ and $k \in \mathbb{N}_0$. Then the following estimates hold:*

$$\|(I-\omega K^*K)^k(K^*K)^s\| \ \leq \ \omega^{-s}(k+1)^{-s}\,, \tag{2.33}$$

$$\|(I-\omega K^*K)^k K^*\| \ \leq \ \omega^{-\frac{1}{2}}(k+1)^{-\frac{1}{2}}\,, \tag{2.34}$$

$$\left\|\omega\sum_{j=0}^{k-1}(I-\omega KK^*)^j(KK^*)^s\right\| \ \leq \ (\omega k)^{1-s}\,. \tag{2.35}$$

Moreover, for any $v \in \mathcal{N}(K)^\perp$ it holds that

$$c_k(s,v) := (k+1)^s\,\|(I-\omega K^*K)^k(K^*K)^s v\| \to 0 \quad as \quad k\to\infty\,. \tag{2.36}$$

Proof. It follows by means of spectral theory (see, e.g., [45]) that

$$\|(I-\omega K^*K)^k(K^*K)^s\| \ \leq \ \omega^{-s}\sup_{\lambda\in[0,1]}(1-\lambda)^k\lambda^s,$$

$$\|(I-\omega K^*K)^k K^*\| \ \leq \ \omega^{-\frac{1}{2}}\sup_{\lambda\in[0,1]}(1-\lambda)^k\lambda^{\frac{1}{2}}\,,$$

$$\left\|\omega\sum_{j=0}^{k-1}(I-\omega KK^*)^j(KK^*)^s\right\| \ \leq \ \omega^{1-s}\sup_{\lambda\in[0,1]}(1-(1-\omega\lambda)^k)\lambda^{s-1}.$$

The estimates now follow by some calculus arguments. To show the last assertion let $\{E_\lambda\}$ be a spectral family of K^*K. Then due to (2.33) and Lebesgue's Dominated Convergence Theorem

$$\begin{aligned}
\lim_{k\to\infty} c_k(s,v)^2 &= \lim_{k\to\infty}(k+1)^{2s}\int_0^{\frac{1}{\omega}+}(1-\omega\lambda)^{2k}\lambda^{2s}\,d\|E_\lambda v\|^2\\
&= \int_0^{\frac{1}{\omega}+}\lim_{k\to\infty}(k+1)^{2s}(1-\omega\lambda)^{2k}\lambda^{2s}\,d\|E_\lambda v\|^2\\
&= \begin{cases} 0\,, & s>0\,,\\ \|Pv\|, & s=0\,,\end{cases}
\end{aligned}$$

where P is the orthogonal projector onto $\mathcal{N}(K)$. Since $v \in \mathcal{N}(K)^\perp$, it follows that $\|Pv\| = 0$. □

Proposition 2.11 *Let Assumption 2.7 hold. If $x^\dagger - x_0$ satisfies* (2.22) *with some $0 < \mu \le 1/2$ and $\|v\|$ sufficiently small, then there exist positive sequences $\gamma_{1,k}$ and $\gamma_{2,k}$, depending on μ and v, with*

$$\begin{aligned} \|x_k^\delta - x^\dagger\| &\le \omega^{-\mu}(k+1)^{-\mu}\gamma_{1,k}\,, & \gamma_{1,k} \le 2\gamma\|v\|\,, \quad (2.37)\\ \|y^\delta - F(x_k^\delta)\| &\le \omega^{-(\mu+\frac{1}{2})}(k+1)^{-(\mu+\frac{1}{2})}\gamma_{2,k}\,, & \gamma_{2,k} \le 8\gamma\|v\|\,, \quad (2.38)\\ \gamma &:= 1 + \frac{\tau}{\tau-2} & (2.39) \end{aligned}$$

for $0 \le k < k_*$. Here k_* is the stopping index of the discrepancy principle (2.2) *with $\tau > 2$. In the case of exact data ($\delta = 0$),* (2.37) *and* (2.38) *hold for all $k \ge 0$. Moreover, $\gamma_{1,k} \to 0$ and $\gamma_{2,k} \to 0$ as $k \to \infty$ if $\mu < 1/2$.*

Proof. To simplify the notation we set $K := F'(x^\dagger)$ and $e_k^\delta := x_k^\delta - x^\dagger$. Note that $e_0^\delta = e_0 = x_0 - x^\dagger$. We will now show by induction that

$$\begin{aligned} \|e_j^\delta\| &\le \omega^{-\mu}(j+1)^{-\mu}\gamma_{1,j}\,, & \gamma_{1,j} \le 2\gamma\|v\|\,, \quad (2.40)\\ \|Ke_j^\delta\| &\le \tfrac{1}{4}\omega^{-(\mu+\frac{1}{2})}(j+1)^{-(\mu+\frac{1}{2})}\gamma_{2,j}\,, & \gamma_{2,j} \le 8\gamma\|v\|\,, \quad (2.41) \end{aligned}$$

holds for all $0 \le j < k_*$ and $\|v\|$ sufficiently small. Especially, we assume that $\|v\|$ is so small that (2.30) and (2.31) hold. This is possible, since $\tau > 2$. Note that, due to (2.22), $\|e_0\| \le \|K\|^{2\mu}\|v\|$.

Now it follows from the proof of Theorem 2.8 that $x_k^\delta \in \mathcal{B}_\rho(x^\dagger) \subset \mathcal{D}(F)$ and that $\|e_k^\delta\|$ is monotonically decreasing for $k \le k_*$.

For $j = 0$ the assertions in (2.40) and (2.41) are trivially satisfied with

$$\gamma_{1,0} = \|v\| \qquad \text{and} \qquad \gamma_{2,0} = 4\|v\|\,,$$

since due to (2.22) and (2.29) the estimates

$$\|e_0\| \le \omega^{-\mu}\|v\| \quad \text{and} \quad \|Ke_0\| \le \omega^{-(\mu+\frac{1}{2})}\|v\|$$

hold.

We now assume that (2.40) and (2.41) are true for all $0 \le j < k$ with some $k \le k_*$. Thus, we obtain from (2.24) and (2.26) the representation

$$\begin{aligned} e_k^\delta = {} & (I - \omega K^*K)e_{k-1}^\delta + \omega\Big[(G^\delta(x^\dagger)^* - K^*)(R_{x_{k-1}^\delta}^\delta)^*(y^\delta - F(x_{k-1}^\delta))\\ & + K^*(y^\delta - y) + K^*((R_{x_{k-1}^\delta}^\delta)^* - I)(y^\delta - F(x_{k-1}^\delta))\\ & - K^*(F(x_{k-1}^\delta) - F(x^\dagger) - Ke_{k-1}^\delta)\Big] \end{aligned}$$

which yields the expression

$$\begin{aligned} e_k^\delta &= (I - \omega K^* K)^k e_0 + \omega \sum_{j=0}^{k-1} (I - \omega K^* K)^{k-j-1} p_j^\delta \\ &\quad + \omega \sum_{j=0}^{k-1} (I - \omega K^* K)^{k-j-1} K^* q_j^\delta \\ &\quad + \omega \Big[\sum_{j=0}^{k-1} (I - \omega K^* K)^j K^* \Big] (y^\delta - y) \end{aligned} \tag{2.42}$$

and consequently

$$\begin{aligned} K e_k^\delta &= (I - \omega K K^*)^k K e_0 + \omega \sum_{j=0}^{k-1} (I - \omega K K^*)^{k-j-1} K p_j^\delta \\ &\quad + \omega \sum_{j=0}^{k-1} (I - \omega K K^*)^{k-j-1} K K^* q_j^\delta \\ &\quad + [I - (I - \omega K K^*)^k](y^\delta - y)\,, \end{aligned} \tag{2.43}$$

where

$$p_j^\delta := (G^\delta(x^\dagger)^* - K^*)(R_{x_j^\delta}^\delta)^*(y^\delta - F(x_j^\delta))\,, \tag{2.44}$$

$$q_j^\delta := ((R_{x_j^\delta}^\delta)^* - I)(y^\delta - F(x_j^\delta)) - (F(x_j^\delta) - F(x^\dagger) - K e_j^\delta)\,. \tag{2.45}$$

We will now derive estimates for $\|e_k^\delta\|$ and $\|K e_k^\delta\|$. For this purpose, we estimate $\|p_j^\delta\|$ and $\|q_j^\delta\|$ with j as above.

First of all note that, by virtue of (1.2), (2.2), and (2.25), we obtain for $j < k \le k_*$ that

$$\begin{aligned} \|y^\delta - F(x_j^\delta)\| &\le \|y^\delta - F(x_j^\delta)\| + \frac{1}{\tau - 1}(\|y^\delta - F(x_j^\delta)\| - \tau\delta) \\ &\le \frac{\tau}{\tau - 1}(\|y^\delta - F(x_j^\delta)\| - \|y^\delta - y\|) \\ &\le 2\|y - F(x_j^\delta)\| \end{aligned}$$

and

$$\|y - F(x_j^\delta)\| \le \frac{1}{1 - c_1 \|e_j^\delta\|} \|K e_j^\delta\|\,.$$

This together with (2.40), the monotonicity of $\|e_j^\delta\|$, and the fact that, due to (2.30), $c_1\|e_0\| \le 1/2$, yields the estimates

$$\|y - F(x_j^\delta)\| \le 2\|Ke_j^\delta\| \quad \text{and} \quad \|y^\delta - F(x_j^\delta)\| \le 4\|Ke_j^\delta\|. \tag{2.46}$$

The conditions (2.25), (2.27), and (2.28) together with (2.44) and (2.45) imply that

$$\begin{aligned}\|p_j^\delta\| &\le c_3\delta(1 + c_2\|e_j^\delta\|)\|y^\delta - F(x_j^\delta)\|,\\ \|q_j^\delta\| &\le \|e_j^\delta\|(c_2\|y^\delta - F(x_j^\delta)\| + c_1\|y - F(x_j^\delta)\|),\end{aligned}$$

which together with (2.40) and (2.46) yields that

$$\|p_j^\delta\| \le c_4\delta\|Ke_j^\delta\|, \qquad c_4 := 4c_3\left(1 + c_2\min\left\{\tfrac{1}{2c_1}, \rho\right\}\right), \tag{2.47}$$

$$\|q_j^\delta\| \le c_5\|e_j^\delta\|\,\|Ke_j^\delta\|, \qquad c_5 := 2c_1 + 4c_2. \tag{2.48}$$

The estimates (2.41) and (2.47) and the proof of Lemma 2.9 ($p = 1/2$ and $q = 0$) imply that

$$\begin{aligned}\Big\|\omega\sum_{j=0}^{k-1}(I - \omega K^*K)^{k-j-1}p_j^\delta\Big\| &\le 2c_4\omega^{\frac{1}{2}-\mu}\gamma\|v\|\delta\sum_{j=0}^{k-1}(j+1)^{-(\mu+\frac{1}{2})}\\ &\le 4c_4\omega^{\frac{1}{2}-\mu}\gamma\|v\|\sqrt{k}\delta\end{aligned}$$

and (2.34), (2.40), (2.41), (2.48), and Lemma 2.9 imply that

$$\begin{aligned}&\Big\|\omega\sum_{j=0}^{k-1}(I - \omega K^*K)^{k-j-1}K^*q_j^\delta\Big\|\\ &\quad\le 4c_5\omega^{-2\mu}\gamma^2\|v\|^2\sum_{j=0}^{k-1}(j+1)^{-(2\mu+\frac{1}{2})}(k-j)^{-\frac{1}{2}}\\ &\quad\le 4c_5\omega^{-2\mu}\gamma^2\|v\|^2(k+1)^{-\mu}c_{6,k}(\mu)\end{aligned}$$

with

$$c_{6,k}(\mu) := c(2\mu + \tfrac{1}{2}, \tfrac{1}{2})(k+1)^{-\mu}\begin{cases}1, & \mu < \frac{1}{4},\\ \ln(k+1), & \mu = \frac{1}{4},\\ (k+1)^{2\mu-\frac{1}{2}}, & \mu > \frac{1}{4},\end{cases} \tag{2.49}$$

$$\le \tfrac{4}{e}c(2\mu + \tfrac{1}{2}, \tfrac{1}{2}). \tag{2.50}$$

Combining the estimates above with (1.2), (2.22), (2.35), (2.36) (with $s = \mu$), and (2.42), we finally get

$$\begin{aligned} \|e_k^\delta\| \;\le\;& (k+1)^{-\mu}(c_k(\mu,v) + 4c_5\omega^{-2\mu}\gamma^2\,\|v\|^2 c_{6,k}(\mu)) \\ &+ (4c_4\omega^{\frac12-\mu}\gamma\,\|v\| + \omega^{\frac12})\sqrt{k}\delta\,. \end{aligned} \tag{2.51}$$

Now we turn to an estimate for $\|Ke_k^\delta\|$. It follows with (2.34), (2.41), (2.47), and Lemma 2.9 that

$$\begin{aligned} &\Big\|\omega\sum_{j=0}^{k-1}(I-\omega KK^*)^{k-j-1}Kp_j^\delta\Big\| \\ &\quad\le\; 2c_4\omega^{-\mu}\gamma\,\|v\|\,\delta\sum_{j=0}^{k-1}(j+1)^{-(\mu+\frac12)}(k-j)^{-\frac12} \\ &\quad\le\; 2c_4c(\tfrac12,\tfrac12)\omega^{-\mu}\gamma\,\|v\|\,\delta \end{aligned} \tag{2.52}$$

and with (2.33), (2.40), (2.41), (2.48), and Lemma 2.9 that

$$\begin{aligned} &\Big\|\omega\sum_{j=0}^{k-1}(I-\omega KK^*)^{k-j-1}KK^*q_j^\delta\Big\| \\ &\quad\le\; 4c_5\omega^{-(2\mu+\frac12)}\gamma^2\,\|v\|^2\sum_{j=0}^{k-1}(j+1)^{-(2\mu+\frac12)}(k-j)^{-1} \\ &\quad\le\; 4c_5\omega^{-(2\mu+\frac12)}\gamma^2\,\|v\|^2(k+1)^{-(\mu+\frac12)}c_{7,k}(\mu) \end{aligned}$$

with

$$c_{7,k}(\mu) \;:=\; c(2\mu+\tfrac12,1)(k+1)^{-\mu}\begin{cases}\ln(k+1)\,, & \mu\le\frac14\,,\\ (k+1)^{2\mu-\frac12}\,, & \mu>\frac14\,,\end{cases} \tag{2.53}$$

$$\le\; \max\Big\{1,\tfrac{1}{\mu e}\Big\}\,c(2\mu+\tfrac12,1)\,. \tag{2.54}$$

A combination of the estimates above with (1.2), (2.22), (2.36) (with $s = \mu+1/2$), and (2.43), now yields that

$$\begin{aligned} \|Ke_k^\delta\| \;\le\;& (k+1)^{-(\mu+\frac12)}(c_k(\mu+\tfrac12,v) \\ &+ 4c_5\omega^{-(2\mu+\frac12)}\gamma^2\,\|v\|^2 c_{7,k}(\mu)) \\ &+ (2c_4c(\tfrac12,\tfrac12)\omega^{-\mu}\gamma\,\|v\| + 1)\delta\,. \end{aligned} \tag{2.55}$$

Note that, due to (2.29), it holds that $\|I-(I-\omega KK^*)^k\|\le 1$.

We now derive an estimate for δ in terms of k. By (1.2), (2.2), (2.25), and the monotonicity of $\|e_k^\delta\|$, we obtain for $k < k_*$ that

$$\tau\delta < \|y^\delta - F(x_k^\delta)\| \leq \delta + \frac{1}{1 - c_1 \|e_0\|} \|K e_k^\delta\| .$$

We will assume that $\|v\|$ is so small that

$$4c_4 \max\{2, c(\tfrac{1}{2}, \tfrac{1}{2})\}\omega^{-\mu}\gamma \|v\| + c_1\tau \|e_0\| \leq \tau - 2 \tag{2.56}$$

holds. Note that $\tau > 2$. Together with (2.55) this yields

$$\delta \leq \frac{2}{(\tau - 2)(1 - c_1 \|e_0\|)} (k+1)^{-(\mu+\frac{1}{2})} c_{8,k}(\mu) \tag{2.57}$$

with

$$c_{8,k}(\mu, v) := c_k(\mu + \tfrac{1}{2}, v) + 4c_5\omega^{-(2\mu+\frac{1}{2})}\gamma^2 \|v\|^2 c_{7,k}(\mu) . \tag{2.58}$$

This together with (2.51), (2.55), and (2.56) yields

$$\|e_k^\delta\| \leq \omega^{-\mu}(k+1)^{-\mu}\gamma_{1,k} \quad \text{and} \quad \|K e_k^\delta\| \leq \tfrac{1}{4}\omega^{-(\mu+\frac{1}{2})}(k+1)^{-(\mu+\frac{1}{2})}\gamma_{2,k}$$

with

$$\gamma_{1,k} := \omega^\mu c_k(\mu, v) + 4c_5\omega^{-\mu}\gamma^2 \|v\|^2 c_{6,k}(\mu) + \frac{\tau}{\tau - 2}\omega^{\mu+\frac{1}{2}} c_{8,k}(\mu, v) , \tag{2.59}$$

$$\gamma_{2,k} := \frac{8(\tau - 1)}{\tau - 2}\omega^{\mu+\frac{1}{2}} c_{8,k}(\mu, v) . \tag{2.60}$$

Together with (2.33), (2.36), (2.39), (2.50), and (2.54) we get the estimate

$$\gamma_{1,k} \leq \gamma \|v\| (1 + c_9\gamma^2 \|v\|) \quad \text{and} \quad \gamma_{2,k} \leq 4\gamma \|v\| (1 + c_9\gamma^2 \|v\|)$$

with

$$c_9 := 4\omega^{-\mu} c_5 \max\left\{\tfrac{4}{e}, \tfrac{1}{\mu e}\right\} \max\{c(2\mu + \tfrac{1}{2}, \tfrac{1}{2}), c(2\mu + \tfrac{1}{2}, 1)\}$$

This shows that (2.40) and (2.41) hold for all $0 \leq j < k_*$ provided $\|v\|$ is sufficiently small, namely so small that (2.30), (2.31), (2.56), and

$$c_9\gamma^2 \|v\| \leq 1$$

hold. Together with (2.46) this shows that the assertions (2.37) and (2.38) are valid.

Let us now assume that $\delta = 0$ and $\mu < 1/2$. Then (2.36), (2.49), (2.53), (2.58), (2.59), and (2.60) imply that $\gamma_{1,k} \to 0$ and $\gamma_{2,k} \to 0$ as $k \to \infty$. □

For the proof of the main result of this section we need the following interesting lemma.

Lemma 2.12 *Let the assumptions of Theorem* 2.8 *hold. If* $k_*(\delta, y^\delta) \not\to \infty$ *as* $\delta \to 0$, *then* $x^\dagger - x_0 \in \mathcal{R}(F'(x^\dagger)^*) = \mathcal{R}((F'(x^\dagger)^* F'(x^\dagger))^{\frac{1}{2}})$.

Proof. As mentioned in the proof of Theorem 2.8 the Landweber iterates depend continuously on y^δ. Thus, $x_k^\delta \to x_k$ as $\delta \to 0$.

Let us now assume that $k_*(\delta, y^\delta) \not\to \infty$ as $\delta \to 0$. Then there exists a sequence $\delta_n \to 0$ as $n \to \infty$ with $k_*(\delta_n, y^{\delta_n}) = \overline{k} \in \mathbb{N}_0$. Using the notation of Proposition 2.11 and the fact that, due to Theorem 2.8, $x_{\overline{k}}^\delta = x_{k_*}^\delta \to x^\dagger$ as $\delta \to 0$, we obtain with (2.42) (with $\delta = 0$)

$$0 = e_{\overline{k}}^0 = (I - \omega K^*K)^{\overline{k}} e_0 + \omega \sum_{j=0}^{\overline{k}-1} (I - \omega K^*K)^{\overline{k}-j-1} K^* q_j^0 .$$

This immediately implies the assertion. □

In the next theorem we prove that for the modified Landweber iteration we obtain the same convergence rates and the same asymptotical estimate for k_* as for linear ill-posed problems (compare [45, Theorem 6.5]) if $\mu \le 1/2$ in (2.22).

Theorem 2.13 *Let Assumption* 2.7 *hold and let* $k_* = k_*(\delta, y^\delta)$ *be chosen according to the stopping rule* (2.2) *with* $\tau > 2$. *If* $x^\dagger - x_0$ *satisfies* (2.22) *with some* $0 < \mu \le 1/2$ *and* $\|v\|$ *sufficiently small, then it holds that*

$$k_* = O\Big(\|v\|^{\frac{2}{2\mu+1}} \delta^{-\frac{2}{2\mu+1}} \Big)$$

and

$$\|x_{k_*}^\delta - x^\dagger\| = \begin{cases} o\Big(\|v\|^{\frac{1}{2\mu+1}} \delta^{\frac{2\mu}{2\mu+1}} \Big), & \mu < \frac{1}{2}, \\ O\Big(\sqrt{\|v\| \delta} \Big), & \mu = \frac{1}{2}. \end{cases}$$

Proof. We use the same notation as in the proof of Proposition 2.11 and assume that $\|v\|$ is so small that all the estimates of that proof hold. If we put

$$f_{k_*} = (I - \omega K^*K)^{k_*} v + \omega \sum_{j=0}^{k_*-1} (I - \omega K^*K)^{k_*-j-1} (K^*K)^{-\mu} K^* q_j^\delta , \quad (2.61)$$

then (2.33), (2.36), (2.40), (2.41), (2.48), and Lemma 2.9 imply the estimate

$$\begin{aligned} \|f_{k_*}\| &\leq c_{k_*}(0,v) + 4c_5\omega^{-\mu}\gamma^2\|v\|^2 \sum_{j=0}^{k_*-1}(j+1)^{-(2\mu+\frac{1}{2})}(k_*-j)^{\mu-\frac{1}{2}} \\ &\leq c_\mu(c_{k_*}(0,v) + c_{1,k_*}\|v\|)\,, \end{aligned} \tag{2.62}$$

where $c_\mu > 0$ plays the role of a generic constant depending on μ and

$$c_{1,k_*} := (k_*+1)^{-\mu}\begin{cases} 1\,, & \mu \leq \frac{1}{4}\,, \\ \ln(k_*+1)\,, & \mu = \frac{1}{4}\,, \\ (k_*+1)^{2\mu-\frac{1}{2}}\,, & \mu > \frac{1}{4}\,. \end{cases} \tag{2.63}$$

On the other hand, we obtain with (1.2), (2.2), (2.22), (2.25), (2.43), (2.52), and (2.61) that

$$\begin{aligned} \|K(K^*K)^\mu f_{k_*}\| &\leq \|Ke_{k_*}^\delta\| + \Big\|\omega\sum_{j=0}^{k_*-1}(I-\omega KK^*)^{k-j-1}Kp_j^\delta\Big\| \\ &\quad + \|[I-(I-\omega KK^*)^k](y^\delta - y)\| \\ &\leq (1+c_1\|e_0\|)\,\|y - F(x_{k_*}^\delta)\| + c_\mu\delta \\ &\leq ((1+c_1\|e_0\|)(1+\tau)+c_\mu)\delta \leq c_\mu\delta\,. \end{aligned}$$

Together with (2.62), the interpolation inequality now yields

$$\|(K^*K)^\mu f_{k_*}\| \leq c_\mu(c_{k_*}(0,v) + c_{1,k_*}\|v\|)^{\frac{1}{2\mu+1}}\delta^{\frac{2\mu}{2\mu+1}}\,.$$

From (2.42), (2.51), and (2.61) we conclude that

$$\|e_{k_*}^\delta\| \leq \|(K^*K)^\mu f_{k_*}\| + c_\mu\sqrt{k_*}\delta\,.$$

If $k_* > 0$, we can apply (2.57) with $k = k_* - 1$ to obtain

$$k_*^{\mu+\frac{1}{2}} \leq c_\mu\delta^{-1}(c_{k_*-1}(\mu+\tfrac{1}{2},v) + c_{2,k_*}\|v\|)\,,$$

where

$$c_{2,k_*} := k_*^{-\mu}\begin{cases} \ln(k_*)\,, & \mu \leq \frac{1}{4}\,, \\ k_*^{2\mu-\frac{1}{2}}\,, & \mu > \frac{1}{4}\,. \end{cases} \tag{2.64}$$

Combining the estimates above we obtain that

$$\|e_{k_*}^\delta\| \leq c_\mu(c_{k_*}(0,v) + c_{k_*-1}(\mu+\tfrac{1}{2},v) + (c_{1,k_*}+c_{2,k_*})\|v\|)^{\frac{1}{2\mu+1}}\delta^{\frac{2\mu}{2\mu+1}}\,. \tag{2.65}$$

Together with (2.33), (2.36), (2.63), and (2.64) this implies already the rate $O\Big(\|v\|^{\frac{1}{2\mu+1}}\delta^{\frac{2\mu}{2\mu+1}}\Big)$.

Let us now assume that $\mu < 1/2$. If $k_*(\delta, y^\delta) \not\to \infty$ as $\delta \to 0$, then it follows with Lemma 2.12 that we could have chosen $\mu = 1/2$ in (2.22) from the very beginning yielding even the better rate $O\Big(\sqrt{\|v\|\,\delta}\Big)$.

If $k_*(\delta, y^\delta) \to \infty$ as $\delta \to 0$, then it follows from (2.36), (2.63), and (2.64) that

$$c_{k_*}(0, v) + c_{k_*-1}(\mu + \tfrac{1}{2}, v) + (c_{1,k_*} + c_{2,k_*})\,\|v\| \to 0 \quad \text{as} \quad \delta \to 0$$

yielding the rate $o\Big(\|v\|^{\frac{1}{2\mu+1}}\delta^{\frac{2\mu}{2\mu+1}}\Big)$. □

Under the Assumption 2.7, and according to Theorem 2.13 the best possible rate of convergence is

$$\|x_{k_*}^\delta - x^\dagger\| = O(\sqrt{\delta})$$

attained when $\mu = 1/2$. Even if $\mu > 1/2$ we cannot improve this rate without an additional restriction of the *nonlinearity* of F. We believe that this is a natural effect as the following argument shows: even without noise we have, due to (2.42), that

$$x^\dagger - x_1 = (K^*K)^\mu (I - \omega K^*K)v - K^* q_0^0\,,$$

and $K^* q_0^0 \notin \mathcal{R}((K^*K)^\mu)$ for any $\mu > 1/2$, in general. Thus, the source condition (2.22) with $\mu > 1/2$ will not remain true for $x^\dagger - x_1$, $x^\dagger - x_2$, and so on, and therefore, we cannot expect a better rate of convergence, in general.

Note that for linear ill-posed problems there is no restriction (cf., e.g., [45]). This can also be seen from the proof of Proposition 2.11, since then $p_j^\delta = q_j^\delta = 0$ and (2.36) holds for all $s \geq 0$.

2.4 An example

In [67] three examples were considered where the conditions for obtaining convergence rates were checked, namely a nonlinear Hammerstein equation and two parameter estimation problems. For the first two examples conditions (2.26) and (2.27) are satisfied with $G^\delta(x) = F'(x)$, i.e., the convergence rates analysis can be applied to the Landweber iterates (2.1). In the third example, where a diffusion coefficient is estimated, this is no longer the case. However, the conditions are satisfied for modified Landweber iterates as in (2.24):

Example 2.14 We treat the problem of estimating the diffusion coefficient a in

$$\begin{aligned} -(a(s)u(s)_s)_s &= f(s)\,, \qquad s \in (0,1)\,, \\ u(0) = 0 &= u(1)\,, \end{aligned} \tag{2.66}$$

where $f \in L^2$; the subscript s denotes derivative with respect to s.

In this example, F is defined as the parameter-to-solution mapping

$$\begin{aligned} F : \mathcal{D}(F) := \{a \in H^1[0,1] \,:\, a(s) \geq \underline{a} > 0\} &\rightarrow L^2[0,1] \\ a &\mapsto F(a) := u(a)\,, \end{aligned}$$

where $u(a)$ is the solution of (2.66). One can prove that F is Fréchet-differentiable with

$$\begin{aligned} F'(a)h &= A(a)^{-1}[(hu_s(a))_s]\,, \\ F'(a)^*w &= -B^{-1}[u_s(a)(A(a)^{-1}w)_s]\,, \end{aligned}$$

where

$$\begin{aligned} A(a) : H^2[0,1] \cap H_0^1[0,1] &\rightarrow L^2[0,1] \\ u &\mapsto A(a)u := -(au_s)_s \end{aligned}$$

and

$$\begin{aligned} B : \mathcal{D}(B) := \{\psi \in H^2[0,1] \,:\, \psi'(0) = \psi'(1) = 0\} &\rightarrow L^2[0,1] \\ \psi &\mapsto B\psi := -\psi'' + \psi\,; \end{aligned}$$

note that B^{-1} is the adjoint of the embedding operator from $H^1[0,1]$ in $L^2[0,1]$.

First of all, we show that F satisfies condition (2.25): let $F(a) = u$, $F(\tilde{a}) = \tilde{u}$, and $w \in L^2$. Noting that $(\tilde{u} - u) \in H^2 \cap H_0^1$ and that $A(a)$ is one-to-one and onto for $a, \tilde{a} \in \mathcal{D}(F)$ we obtain that

$$\begin{aligned} &\langle\, F(\tilde{a}) - F(a) - F'(a)(\tilde{a} - a), w \,\rangle_{L^2} \\ &\quad = \langle\, (\tilde{u} - u) - A(a)^{-1}[((\tilde{a} - a)u_s)_s], w \,\rangle_{L^2} \\ &\quad = \langle\, A(a)(\tilde{u} - u) - ((\tilde{a} - a)u_s)_s, A(a)^{-1}w \,\rangle_{L^2} \\ &\quad = \langle\, ((\tilde{a} - a)(\tilde{u}_s - u_s))_s, A(a)^{-1}w \,\rangle_{L^2} \\ &\quad = -\langle\, (\tilde{a} - a)(\tilde{u} - u)_s, (A(a)^{-1}w)_s \,\rangle_{L^2} \\ &\quad = \langle\, F(\tilde{a}) - F(a), ((\tilde{a} - a)(A(a)^{-1}w)_s)_s \,\rangle_{L^2}\,. \end{aligned}$$

This together with the fact that $\|g\|_{L^\infty} \leq \sqrt{2}\|g\|_{H^1}$ and that $\|g\|_{L^\infty} \leq \|g'\|_{L^2}$ if $g \in H^1$ is such that $g(\xi) = 0$ for some $\xi \in [0,1]$, yields the estimate

$$\begin{aligned} &\|F(\tilde{a}) - F(a) - F'(a)(\tilde{a} - a)\|_{L^2} \\ &\quad \leq \sup_{\|w\|_{L^2}=1} \langle\, F(\tilde{a}) - F(a), ((\tilde{a} - a)(A(a)^{-1}w)_s)_s \,\rangle_{L^2} \end{aligned}$$

$$\le \|F(\tilde{a}) - F(a)\|_{L^2} \sup_{\|w\|_{L^2}=1} \Big[\Big\| \Big(\frac{\tilde{a}-a}{a} \Big)_s \Big\|_{L^2} \|a(A(a)^{-1}w)_s\|_{L^\infty}$$

$$+ \Big\| \frac{\tilde{a}-a}{a} \Big\|_{L^\infty} \|w\|_{L^2} \Big]$$

$$\le \underline{a}^{-1}(1+\sqrt{2}+\underline{a}^{-1}\sqrt{2}\|a\|_{H^1})\,\|F(\tilde{a})-F(a)\|_{L^2}\,\|\tilde{a}-a\|_{H^1}\,. \quad (2.67)$$

This implies (2.25).

As mentioned in the introduction of this section, the conditions (2.26) and (2.27) are not fulfilled with $G^\delta(x) = F'(x)$. Noting that $F'(a)^*w$ is the unique solution of the variational problem: for all $v \in H^1$

$$\langle (F'(a)^*w)_s, v_s \rangle_{L^2} + \langle F'(a)^*w, v \rangle_{L^2} = \langle u(a), ((A(a)^{-1}w)_s v)_s \rangle_{L^2}\,, \quad (2.68)$$

we propose to choose G^δ in (2.24) as follows: $G^\delta(a)^*w = G(a,u^\delta)^*w$ is the unique solution g of the variational problem

$$\langle g_s, v_s \rangle_{L^2} + \langle g, v \rangle_{L^2} = \langle u^\delta, ((A(a)^{-1}w)_s v)_s \rangle_{L^2}\,, \quad v \in H^1\,. \quad (2.69)$$

This operator G^δ obviously satisfies (2.26), since

$$G(\tilde{a},u^\delta)^* = G(a,u^\delta)^* R(\tilde{a},a)^*$$

with

$$R(\tilde{a},a)^* = A(a)A(\tilde{a})^{-1}.$$

The condition (2.27) is satisfied, since one can estimate as in (2.67) that

$$\|R(\tilde{a},a)^* - I\| = \|A(a)A(\tilde{a})^{-1} - I\| \le \underline{a}^{-1}(1+\sqrt{2}+\underline{a}^{-1}\sqrt{2}\|\tilde{a}\|_{H^1})\,\|\tilde{a}-a\|_{H^1}\,.$$

Note that a constant c_2 independent from $\tilde{a}$ can be found, since it is assumed that $\tilde{a} \in \mathcal{B}_\rho(a)$. Now we turn to condition (2.28): using (2.68) and (2.69) we obtain similarly to (2.67) the estimate

$$\|(F'(a)^* - G(a,u^\delta)^*)w\|_{H^1} = \sup_{\|v\|_{H^1}=1} \langle u(a)-u^\delta, ((A(a)^{-1}w)_s v)_s \rangle_{L^2}$$

$$\le \underline{a}^{-1}(1+\sqrt{2}+\underline{a}^{-1}\sqrt{2}\|a\|_{H^1})\,\|u(a)-u^\delta\|_{L^2}\,\|w\|_{L^2}\,.$$

This together with $F(a^\dagger) = u(a^\dagger)$ and $\|u^\delta - u(a^\dagger)\|_{L^2} \le \delta$ implies that

$$\|F'(a^\dagger) - G(a^\dagger,u^\delta)\| \le \underline{a}^{-1}(1+\sqrt{2}+\underline{a}^{-1}\sqrt{2}\|a^\dagger\|_{H^1})\,\delta$$

and hence (2.28) holds.

Thus, Theorem 2.13 is applicable, i.e., if ω and τ are chosen appropriately, then the modified Landweber iterates $a^\delta_{k_*}$ (cf. (2.24)) where k_* is chosen according to the stopping rule (2.2) converge to the exact solution $a^\dagger$ with the rate $O(\sqrt{\delta})$ provided that

$$a^\dagger - a_0 = -B^{-1}[u_s(a)(A(a)^{-1}w)_s]$$

with $\|w\|$ sufficiently small. Note that this means that

$$a^\dagger - a_o \in H^3\,, \qquad (a^\dagger - a_0)_s(0) = 0 = (a^\dagger - a_0)_s(1)\,,$$

$$z := \frac{(a^\dagger - a_0)_{ss} - (a^\dagger - a_0)}{u_s(a)} \in H^1\,, \qquad \int_0^1 z(s)\,ds = 0\,.$$

Basically this means that one has to know all rough parts of $a^\dagger$ up to H^3. But without this knowledge one cannot expect to get the rate $O(\sqrt{\delta})$.

3 Modified Landweber methods

In this chapter we deal with some modifications of the Landweber iteration. In the first section we show that better rates may be obtained for solutions that satisfy stronger smoothness conditions if the iteration is performed in a subspace of $\mathcal{X}$ with a stronger norm. This leads us directly to regularization in Hilbert scales. On the other hand for solutions with poor smoothness properties the number of iterations may be reduced if the iteration is performed in a space with a weaker norm.

By adding an additional penalty term to the iteration scheme, one can obtain convergence rates results under weaker restrictions on the nonlinearity of F (see Section 3.2). A variant of this approach that does not need derivatives of F is mentioned in Section 3.3.

In Section 3.4 we prove convergence of the steepest descent and the minimal error method. These methods can be viewed as Landweber iteration with variable coefficients ω. Finally, in the last section of this chapter we study the Landweber–Kaczmarz method, a variant of Landweber iteration for problems where F describes a system of equations.

3.1 Landweber iteration in Hilbert scales

It is well known for Tikhonov regularization that convergence rates can be improved if the exact solution is smooth enough and if the regularizing norm in $\mathcal{X}$ is replaced by a stronger one in a Hilbert scale (cf. [45, 120, 123]). The same is true for Landweber iteration (cf. [124]). Before we can go into details of this modification, we shortly repeat the definition of a Hilbert scale:

Let L be a densely defined unbounded selfadjoint strictly positive operator in $\mathcal{X}$. Then $(\mathcal{X}_s)_{s\in\mathbb{R}}$ denotes the Hilbert scale induced by L if $\mathcal{X}_s$ is the completion of $\bigcap_{k=0}^{\infty} D(L^k)$ with respect to the Hilbert space norm $\|x\|_s := \|L^s x\|_{\mathcal{X}}$; obviously $\|x\|_0 = \|x\|_{\mathcal{X}}$ (see [102] or [45, Section 8.4] for details).

We will now replace $F'(x_k^\delta)^*$ in (2.1) by the adjoint of $F'(x_k^\delta)$ considered as an operator from $\mathcal{X}_s$ into $\mathcal{Y}$. Usually $s \geq 0$, but we will see below that there are special cases where a negative choice of s can be advantageous. Since by definition of $\mathcal{X}_s$ this adjoint is given by $L^{-2s}F'(x_k^\delta)^*$, (2.1) is replaced by the iteration process

$$x_{k+1}^\delta = x_k^\delta + L^{-2s}F'(x_k^\delta)^*(y^\delta - F(x_k^\delta)), \qquad k \in \mathbb{N}_0 . \tag{3.1}$$

As in the previous chapter the iteration process is stopped according to the discrepancy principle (2.2). As always we assume that the data y^δ satisfy (1.2), i.e., $\|y^\delta - y\| \le \delta$, and that $x_0^\delta = x_0$ is an initial guess.

To prove convergence rates we need some basic conditions. We will relax the conditions used in [124] as suggested in [42]. For an approach, where the Hilbert scale is chosen in the space Y, see [41].

Assumption 3.1

(i) $F : \mathcal{D}(F)(\subset \mathcal{X}) \to \mathcal{Y}$ is continuous and Fréchet-differentiable in $\mathcal{X}$.

(ii) $F(x) = y$ has a solution $x^\dagger$.

(iii) $\|F'(x^\dagger)x\| \le \overline{m}\|x\|_{-a}$ for all $x \in \mathcal{X}$ and some $a > 0$, $\overline{m} > 0$. Moreover, the extension of $F'(x^\dagger)$ to $\mathcal{X}_{-a}$ is injective.

(iv) $B := F'(x^\dagger)L^{-s}$ is such that $\|B\|_{\mathcal{X},\mathcal{Y}} \le 1$, where $-a < s$. If $s < 0$, $F'(x^\dagger)$ has to be replaced by its extension to $\mathcal{X}_s$.

Usually, for the analysis of regularization methods in Hilbert scales a stronger condition than (iii) is used, namely (cf., e.g., [124])

$$\|F'(x^\dagger)x\| \sim \|x\|_{-a} \qquad \text{for all } x \in \mathcal{X}, \tag{3.2}$$

where the number a can be interpreted as a *degree of ill-posedness* of the linearized problem in $x^\dagger$. However, this condition is not always fulfilled. Sometimes one can only prove that condition (iii) in Assumption 3.1 holds. It might also be possible that one can prove an estimate from below in a slightly weaker norm (see examples in [42]), i.e.,

$$\|F'(x^\dagger)x\| \ge \underline{m}\|x\|_{-\tilde{a}} \qquad \text{for all } x \in \mathcal{X} \text{ and some } \tilde{a} \ge a, \underline{m} > 0. \tag{3.3}$$

The next proposition will shed more light onto condition (iii) in Assumption 3.1 and (3.3).

Proposition 3.2 *Let Assumption* 3.1 *hold. Then for all* $\nu \in [0, 1]$ *it holds that*

$$\mathcal{D}((B^*B)^{-\frac{\nu}{2}}) = \mathcal{R}((B^*B)^{\frac{\nu}{2}}) \subset \mathcal{X}_{\nu(a+s)},$$

$$\|(B^*B)^{\frac{\nu}{2}}x\| \le \overline{m}^{\nu}\|x\|_{-\nu(a+s)} \qquad \textit{for all } x \in \mathcal{X}, \tag{3.4}$$

$$\|(B^*B)^{-\frac{\nu}{2}}x\| \ge \overline{m}^{-\nu}\|x\|_{\nu(a+s)} \qquad \textit{for all } x \in \mathcal{D}((B^*B)^{-\frac{\nu}{2}}). \tag{3.5}$$

Note that condition (iii) is equivalent to

$$\mathcal{R}(F'(x^\dagger)^*) \subset \mathcal{X}_a \quad \textit{and} \quad \|F'(x^\dagger)^*w\|_a \le \overline{m}\|w\| \quad \textit{for all } w \in \mathcal{Y}. \tag{3.6}$$

If in addition condition (3.3) *holds, then for all* $\nu \in [0,1]$ *it holds that*

$$\mathcal{X}_{\nu(\tilde{a}+s)} \subset \mathcal{R}((B^*B)^{\frac{\nu}{2}}) = \mathcal{D}((B^*B)^{-\frac{\nu}{2}}),$$

$$\|(B^*B)^{\frac{\nu}{2}}x\| \geq \underline{m}^{\nu}\|x\|_{-\nu(\tilde{a}+s)} \quad \text{for all} \quad x \in \mathcal{X}, \tag{3.7}$$

$$\|(B^*B)^{-\frac{\nu}{2}}x\| \leq \underline{m}^{-\nu}\|x\|_{\nu(\tilde{a}+s)} \quad \text{for all} \quad x \in \mathcal{X}_{\nu(\tilde{a}+s)}. \tag{3.8}$$

Note that condition (3.3) *is equivalent to*

$$\mathcal{X}_{\tilde{a}} \subset \mathcal{R}(F'(x^\dagger)^*) \quad \text{and} \quad \|F'(x^\dagger)^*w\|_{\tilde{a}} \geq \underline{m}\|w\| \\ \text{for all} \quad w \in \mathcal{N}(F'(x^\dagger)^*)^\perp \text{ with } F'(x^\dagger)^*w \in \mathcal{X}_{\tilde{a}}. \tag{3.9}$$

Proof. The proof follows the lines of Corollary 8.22 in [45] noting that the results there not only hold for $s \geq 0$ but also for $s > -a$. □

In our convergence analysis the following *shifted* Hilbert scale will play an important role

$$\widetilde{\mathcal{X}}_r := \mathcal{D}((B^*B)^{\frac{s-r}{2(a+s)}}L^s) \quad \text{equipped with the norm} \\ \|\!|x|\!\|_r := \|(B^*B)^{\frac{s-r}{2(a+s)}}L^s x\|_{\mathcal{X}}, \tag{3.10}$$

where a, s, and B are as in Assumption 3.1. In the next proposition we summarize some properties of this shifted Hilbert scale.

Proposition 3.3 *Let Assumption* 3.1 *hold and let* $(\widetilde{\mathcal{X}}_r)_{r\in\mathbb{R}}$ *be defined as in* (3.10).

(i) The space $\widetilde{\mathcal{X}}_q$ *is continuously embedded in* $\widetilde{\mathcal{X}}_p$ *for* $p < q$*, i.e., for* $x \in \widetilde{\mathcal{X}}_q$

$$\|\!|x|\!\|_p \leq \gamma^{p-q}\|\!|x|\!\|_q, \tag{3.11}$$

where γ *is such that*

$$\langle (B^*B)^{-\frac{1}{2(a+s)}}x, x\rangle \geq \gamma\|x\|^2 \quad \text{for all} \quad x \in \mathcal{D}((B^*B)^{-\frac{1}{2(a+s)}}).$$

(ii) The interpolation inequality holds, i.e., for all $x \in \widetilde{\mathcal{X}}_r$

$$\|\!|x|\!\|_q \leq \|\!|x|\!\|_p^{\frac{r-q}{r-p}}\|\!|x|\!\|_r^{\frac{q-p}{r-p}}, \qquad p < q < r. \tag{3.12}$$

(iii) The following estimates hold:

$$\|x\|_r \leq \overline{m}^{\frac{r-s}{a+s}}\|\!|x|\!\|_r \quad \text{for all } x \in \widetilde{\mathcal{X}}_r \subset \mathcal{X}_r \text{ if } s \leq r \leq a+2s, \\ \|x\|_r \geq \overline{m}^{\frac{r-s}{a+s}}\|\!|x|\!\|_r \quad \text{for all } x \in \mathcal{X}_r \subset \widetilde{\mathcal{X}}_r \text{ if } -a \leq r \leq s. \tag{3.13}$$

Especially, we obtain the estimate

$$\|x\|_0 \le \overline{m}^{\frac{-s}{a+s}} |||x|||_0 \quad \textit{for all} \;\; x \in \widetilde{\mathcal{X}}_0 \subset \mathcal{X}_0 \quad \textit{if} \quad s \le 0\,. \tag{3.14}$$

Moreover,

$$|||x|||_{-a} = \|F'(x^\dagger)x\| \quad \textit{for all} \;\; x \in \mathcal{X}\,. \tag{3.15}$$

(iv) If in addition (3.3) *holds, the following estimates hold with* $p = s + \frac{r-s}{a+s}(\tilde{a} + s)$ *:*

$$\begin{aligned} \|x\|_p &\ge \underline{m}^{\frac{r-s}{a+s}} |||x|||_r \quad \textit{for all } x \in \mathcal{X}_p \subset \widetilde{\mathcal{X}}_r \textit{ if } s \le r \le a + 2s\,, \\ \|x\|_p &\le \underline{m}^{\frac{r-s}{a+s}} |||x|||_r \quad \textit{for all } x \in \widetilde{\mathcal{X}}_r \subset \mathcal{X}_p \textit{ if } -a \le r \le s\,. \end{aligned} \tag{3.16}$$

Especially, we obtain the estimate

$$\|x\|_0 \le \underline{m}^{\frac{-s}{a+s}} |||x|||_r \quad \textit{for all } x \in \widetilde{\mathcal{X}}_r \subset \mathcal{X}_0\,,\; r = s\tfrac{\tilde{a}-a}{\tilde{a}+s}\,,\; \textit{if } s > 0\,. \tag{3.17}$$

If $\tilde{a} = a$*, i.e., in case* (3.2) *holds,*

$$\|x\|_r \sim |||x|||_r \quad \textit{for all } x \in \mathcal{X}_r = \widetilde{\mathcal{X}}_r \textit{ if } -a \le r \le a + 2s\,. \tag{3.18}$$

Proof. The proof follows from Proposition 8.19 in [45] and Proposition 3.2. □

For the convergence rates analysis we need some smoothness conditions on the solution $x^\dagger$ and the Fréchet-derivative of F.

Assumption 3.4

(i) $x_0 \in \widetilde{\mathcal{B}}_\rho(x^\dagger) := \{x \in \mathcal{X} \,:\, x - x^\dagger \in \widetilde{\mathcal{X}}_0 \wedge |||x - x^\dagger|||_0 \le \rho\} \subset \mathcal{D}(F)$ for some $\rho > 0$.

(ii) $\|F'(x^\dagger) - F'(x)\|_{\widetilde{\mathcal{X}}_{-b},\mathcal{Y}} \le c|||x^\dagger - x|||_0^\beta$ for all $x \in \widetilde{\mathcal{B}}_\rho(x^\dagger)$ and some $b \in [0, a]$, $\beta \in (0, 1]$, and $c > 0$.

(iii) $x^\dagger - x_0 \in \widetilde{\mathcal{X}}_u$ for some $(a - b)/\beta < u \le b + 2s$, i.e., there is an element $v \in \mathcal{X}$ so that

$$L^s(x^\dagger - x_0) = (B^*B)^{\frac{u-s}{2(a+s)}} v \quad \text{and} \quad \|v\|_0 = |||x_0 - x^\dagger|||_u\,. \tag{3.19}$$

Before we start our analysis we want to discuss the conditions above.

Remark 3.5 First of all we want to mention that, if (3.2) holds, then, due to (3.18), the conditions in Assumptions 3.1 and 3.4 are equivalent to the ones in Assumption 2.1 in [124].

Note that, due to (3.15), $F'(x^\dagger)$ has a continuous extension to $\widetilde{\mathcal{X}}_{-a} \supset \mathcal{X}_{-a}$. Therefore, condition (ii) in Assumption 3.4 implies that $F'(x)$ has at least a continuous extension to $\widetilde{\mathcal{X}}_{-b} \supset \mathcal{X}$ in a neighbourhood of $x^\dagger$. By definition of the space $\widetilde{\mathcal{X}}_{-b}$, condition (ii) is equivalent to

$$\|(B^*B)^{-\frac{b+s}{2(a+s)}} L^{-s}(F'(x^\dagger)^* - F'(x)^*)\|_{\mathcal{Y},\mathcal{X}} \le c\|x^\dagger - x\|_0^\beta . \tag{3.20}$$

By virtue of (3.6) and Proposition 3.3 (iii), this implies that $L^{-2s}F'(x_k^\delta)^*$ maps $\mathcal{Y}$ at least into $\widetilde{\mathcal{X}}_{b+2s} \subset \mathcal{X}_{b+2s}$ and hence $F'(x_k^\delta)^*$ maps $\mathcal{Y}$ at least into $\mathcal{X}_b$ while $F'(x^\dagger)^*$ maps $\mathcal{Y}$ into $\mathcal{X}_a$.

Note that, if $s = 0$, (3.20) reduces to

$$\|(F'(x^\dagger)^*F'(x^\dagger))^{-\frac{b}{2a}}(F'(x^\dagger)^* - F'(x)^*)\|_{\mathcal{Y},\mathcal{X}} \le c\|x^\dagger - x\|_0^\beta$$

(compare [67, (3.18)]). Moreover, if $b = a$ and $\beta = 1$, this condition is equivalent to (2.27) with $\|R_x^\delta - I\|$ replaced by $\|(R_x^\delta - I)Q\|$, where Q is the orthogonal projector from $\mathcal{Y}$ onto $\overline{\mathcal{R}(F'(x^\dagger))}$.

Condition (iii) in Assumption 3.4 is a smoothness condition for the exact solution comparable to (2.22). Usually $\mathcal{X}_u$ is used instead of $\widetilde{\mathcal{X}}_u$. However, as mentioned above, these conditions are equivalent if (3.2) holds. If $b = a$, then $u \le a + 2s$ is allowed, which is the usual restriction for regularization in Hilbert scales. Note that for Tikhonov regularization of nonlinear problems in Hilbert scales we needed the restriction $a \le s \le u$ (cf. [123]). However, one can show that under the stronger Lipschitz condition (3.20) (with $\beta = 1$), the results in [123] are even valid for $a - b \le s \le u$. For Landweber iteration even oversmoothing (i.e., $u < s$) is allowed as for Tikhonov regularization in the linear case (cf. [45, Section 8.5]).

Lemma 3.6 *Let Assumptions* 3.1 *and* 3.4 *hold. Moreover, let* $k_* = k_*(\delta, y^\delta)$ *be chosen according to the stopping rule* (2.2) *with* $\tau > 2$, *and assume that* $\|e_j^\delta\|_0 \le \rho$ *and that* $\|e_j^\delta\|_u \le \rho_u$ *for all* $0 \le j < k \le k_*$ *and some* $\rho_u > 0$, *where* $e_j^\delta := x_j^\delta - x^\dagger$. *Then there is a positive constant* γ_1 *(independent of* k*) so that the following estimates hold:*

$$\begin{aligned}\|e_k^\delta\|_u \;&\le\; \|x_0 - x^\dagger\|_u + \delta k^{\frac{a+u}{2(a+s)}} \\ &\quad + \gamma_1 \sum_{j=0}^{k-1} (k-j)^{-\frac{a+2s-u}{2(a+s)}} \|e_j^\delta\|_{-a}^{\frac{b+u(1+\beta)}{a+u}} \|e_j^\delta\|_u^{\frac{a(1+\beta)-b}{a+u}}\end{aligned} \tag{3.21}$$

$$
\begin{aligned}
& \quad + \gamma_1 \sum_{j=0}^{k-1} (k-j)^{-\frac{b+2s-u}{2(a+s)}} \|e_j^\delta\|_{-a}^{\frac{a+u(1+\beta)}{a+u}} \|e_j^\delta\|_u^{\frac{a\beta}{a+u}}, \\
\|e_k^\delta\|_{-a} \;\le\;& (k+1)^{-\frac{a+u}{2(a+s)}} \|x_0 - x^\dagger\|_u + \delta \\
& + \gamma_1 \sum_{j=0}^{k-1} (k-j)^{-1} \|e_j^\delta\|_{-a}^{\frac{b+u(1+\beta)}{a+u}} \|e_j^\delta\|_u^{\frac{a(1+\beta)-b}{a+u}} \qquad (3.22) \\
& + \gamma_1 \sum_{j=0}^{k-1} (k-j)^{-\frac{b+a+2s}{2(a+s)}} \|e_j^\delta\|_{-a}^{\frac{a+u(1+\beta)}{a+u}} \|e_j^\delta\|_u^{\frac{a\beta}{a+u}}.
\end{aligned}
$$

Proof. From (3.1) we immediately obtain the representation

$$
e_{k+1}^\delta = (I - L^{-2s} F'(x^\dagger)^* F'(x^\dagger)) e_k^\delta + L^{-2s} F'(x^\dagger)^* (y^\delta - y - q_k^\delta) + L^{-2s} p_k^\delta
$$

with

$$
\begin{aligned}
q_k^\delta &:= F(x_k^\delta) - F(x^\dagger) - F'(x^\dagger) e_k^\delta, && (3.23) \\
p_k^\delta &:= (F'(x_k^\delta)^* - F'(x^\dagger)^*)(y^\delta - F(x_k^\delta)), && (3.24)
\end{aligned}
$$

and furthermore, due to the definition of B (cf. Assumption 3.1 (iv)), the closed expression (note that $e_0^\delta = e_0$)

$$
e_k^\delta = L^{-s}(I - B^*B)^k L^s e_0 + \sum_{j=0}^{k-1} L^{-s}(I - B^*B)^{k-j-1}(B^*(y^\delta - y - q_j^\delta) + L^{-s} p_j^\delta).
$$

Together with (1.2), (3.10), and (3.19) we now obtain the estimates

$$
\begin{aligned}
\|e_k^\delta\|_u \;\le\;& \|(B^*B)^{\frac{s-u}{2(a+s)}} (I - B^*B)^k (B^*B)^{\frac{u-s}{2(a+s)}} v\|_0 \\
& + \|(B^*B)^{\frac{s-u}{2(a+s)}} \sum_{j=0}^{k-1} (I - B^*B)^j B^*\| \,\delta \\
& + \sum_{j=0}^{k-1} \|(B^*B)^{\frac{s-u}{2(a+s)}} (I - B^*B)^{k-j-1} B^*\| \, \|q_j^\delta\| \qquad (3.25) \\
& + \sum_{j=0}^{k-1} \|(B^*B)^{\frac{s-u}{2(a+s)}} (I - B^*B)^{k-j-1} (B^*B)^{\frac{b+s}{2(a+s)}}\| \\
& \quad \cdot \|(B^*B)^{-\frac{b+s}{2(a+s)}} L^{-s} p_j^\delta\|_0
\end{aligned}
$$

and

$$\begin{aligned}
\|e_k^\delta\|_{-a} &\leq \|(B^*B)^{\frac{1}{2}}(I - B^*B)^k (B^*B)^{\frac{u-s}{2(a+s)}} v\|_0 \\
&\quad + \|(B^*B)^{\frac{1}{2}} \sum_{j=0}^{k-1} (I - B^*B)^j B^*\| \, \delta \\
&\quad + \sum_{j=0}^{k-1} \|(B^*B)^{\frac{1}{2}}(I - B^*B)^{k-j-1} B^*\| \, \|q_j^\delta\| \qquad (3.26)\\
&\quad + \sum_{j=0}^{k-1} \|(B^*B)^{\frac{1}{2}}(I - B^*B)^{k-j-1} (B^*B)^{\frac{b+s}{2(a+s)}}\| \\
&\qquad \cdot \|(B^*B)^{-\frac{b+s}{2(a+s)}} L^{-s} p_j^\delta\|_0 .
\end{aligned}$$

Next we derive estimates for $\|q_j^\delta\|$ and $\|(B^*B)^{-\frac{b+s}{2(a+s)}} L^{-s} p_j^\delta\|_0$. Assumption 3.4 (ii), (3.12), and (3.23) imply that

$$\begin{aligned}
\|q_j^\delta\| &\leq \int_0^1 \|F'(x_j^\delta + \xi(x^\dagger - x_j^\delta)) - F'(x^\dagger)) e_j^\delta\| \, d\xi \\
&\leq \tfrac{c}{\beta+1} \|e_j^\delta\|_{-b} \|e_j^\delta\|_0^\beta \leq \tfrac{c}{\beta+1} \|e_j^\delta\|_{-a}^{\frac{b+u(1+\beta)}{a+u}} \|e_j^\delta\|_u^{\frac{a(1+\beta)-b}{a+u}} . \qquad (3.27)
\end{aligned}$$

Since $\tau > 2$, (1.2) and (2.2) imply that for all $0 \leq k < k_*$

$$\begin{aligned}
\|y^\delta - F(x_k^\delta)\| &< \|y^\delta - F(x_k^\delta)\| + (\|y^\delta - F(x_k^\delta)\| - 2\delta) \\
&\leq 2(\|y^\delta - F(x_k^\delta)\| - \|y - y^\delta\|) \leq 2\|y - F(x_k^\delta)\| .
\end{aligned}$$

Thus, we obtain together with (3.15), (3.20), (3.23), (3.24), and $F(x^\dagger) = y$ (cf. Assumption 3.1 (ii)) that

$$\begin{aligned}
\|(B^*B)^{-\frac{b+s}{2(a+s)}} L^{-s} p_j^\delta\|_0 &\leq 2c\|y - F(x_j^\delta)\| \, \|e_j^\delta\|_0^\beta \\
&\leq 2c(\|q_j^\delta\| + \|e_j^\delta\|_{-a}) \|e_j^\delta\|_0^\beta
\end{aligned} \qquad (3.28)$$

for all $0 \leq j < k$.

Since, due to (3.11) and (3.27),

$$\|q_j^\delta\| \leq \tfrac{c}{\beta+1} \gamma^{a-b-u\beta} \|e_j^\delta\|_{-a} \|e_j^\delta\|_u^\beta , \qquad (3.29)$$

(3.12), (3.28), and the assumption that $\|e_j^\delta\|_u \leq \rho_u$ now imply that

$$\|(B^*B)^{-\frac{b+s}{2(a+s)}} L^{-s} p_j^\delta\|_0 \leq \tilde{\gamma} \|e_j^\delta\|_{-a}^{\frac{a+u(1+\beta)}{a+u}} \|e_j^\delta\|_u^{\frac{a\beta}{a+u}} \qquad (3.30)$$

holds for all $0 \leq j < k$ and some $\tilde{\gamma} > 0$ (independent of k).

Combining the estimates (3.25), (3.26), (3.27), and (3.30), and using Lemma 2.10 and (3.19), now yield the assertions (3.21) and (3.22). □

We are now in the position to state the main results of this section:

Proposition 3.7 *Let Assumptions* 3.1 *and* 3.4 *hold. Moreover, let* $k_* = k_*(\delta, y^\delta)$ *be chosen according to the stopping rule* (2.2) *with* $\tau > 2$ *and let* $\|x_0 - x^\dagger\|_u$ *be sufficiently small. Then*

$$\|x_k^\delta - x^\dagger\|_r \le \tfrac{4(\tau-1)}{\tau-2}\|x_0 - x^\dagger\|_u (k+1)^{-\frac{u-r}{2(a+s)}} \tag{3.31}$$

for $-a \le r < u$ *and*

$$\|y^\delta - F(x_k^\delta)\| \le \tfrac{2\tau^2}{\tau-2}\|x_0 - x^\dagger\|_u (k+1)^{-\frac{a+u}{2(a+s)}} \tag{3.32}$$

for all $0 \le k < k_*$. *In the case of exact data* $(\delta = 0)$, *the estimates above hold for all* $k \ge 0$.

Proof. We proceed similar as in the proof of Proposition 2.11. As a first step we show by induction that

$$\|e_j^\delta\|_u \le \eta\|x_0 - x^\dagger\|_u\,, \qquad 0 \le j \le k_*\,, \tag{3.33}$$

and

$$\|e_j^\delta\|_{-a} \le \eta(j+1)^{-\frac{a+u}{2(a+s)}}\|x_0 - x^\dagger\|_u\,, \qquad 0 \le j < k_*\,, \tag{3.34}$$

hold if $\|x_0 - x^\dagger\|_u$ is sufficiently small and

$$\eta = \frac{4(\tau-1)}{\tau-2}\,. \tag{3.35}$$

As mentioned in Remark 3.5 $x_j^\delta - x_0 \in \mathcal{X}_{b+2s} \subset \mathcal{X}$, hence $x_j^\delta \in \mathcal{X}$. Therefore, (3.11) and (3.33) imply that x_j^δ remains in $\widetilde{\mathcal{B}}_\rho(x^\dagger)$ if $\|x_0 - x^\dagger\|_u$ is so small that $\gamma^{-u}\eta\|x_0 - x^\dagger\|_u \le \rho$. We will assume this in the following. Note that this guarantees that the iteration process (3.1) is well defined.

By virtue of (3.21) and (3.22), (3.33) and (3.34) are true for $j = 0$; note that, due to (3.26), estimate (3.22) is valid without the term δ for $k = 0$.

Let us now assume that (3.33) and (3.34) are true for all $0 \le j < k < k_*$. Then we have to verify (3.33) and (3.34) for $j = k$:

Using (3.21) and (3.22) as well as (3.33) and (3.34) for $j < k$ it follows that

$$\begin{aligned}
\|e_k^\delta\|_u &\leq \|x_0 - x^\dagger\|_u + \delta k^{\frac{a+u}{2(a+s)}} \\
&+ \gamma_1(\eta\|x_0 - x^\dagger\|_u)^{1+\beta} \sum_{j=0}^{k-1} (k-j)^{-\frac{a+2s-u}{2(a+s)}} (j+1)^{-\frac{b+u(1+\beta)}{2(a+s)}} \\
&+ \gamma_1(\eta\|x_0 - x^\dagger\|_u)^{1+\beta} \sum_{j=0}^{k-1} (k-j)^{-\frac{b+2s-u}{2(a+s)}} (j+1)^{-\frac{a+u(1+\beta)}{2(a+s)}}, \\
\|e_k^\delta\|_{-a} &\leq (k+1)^{-\frac{a+u}{2(a+s)}} \|x_0 - x^\dagger\|_u + \delta \\
&+ \gamma_1(\eta\|x_0 - x^\dagger\|_u)^{1+\beta} \sum_{j=0}^{k-1} (k-j)^{-1} (j+1)^{-\frac{b+u(1+\beta)}{2(a+s)}} \\
&+ \gamma_1(\eta\|x_0 - x^\dagger\|_u)^{1+\beta} \sum_{j=0}^{k-1} (k-j)^{-\frac{b+a+2s}{2(a+s)}} (j+1)^{-\frac{a+u(1+\beta)}{2(a+s)}}.
\end{aligned}$$

Furthermore with Lemma 2.9 the following estimates,

$$\|e_k^\delta\|_u \leq \|x_0 - x^\dagger\|_u (1 + \gamma_2 \|x_0 - x^\dagger\|_u^\beta) + \delta k^{\frac{a+u}{2(a+s)}}, \tag{3.36}$$

$$\|e_k^\delta\|_{-a} \leq (k+1)^{-\frac{a+u}{2(a+s)}} \|x_0 - x^\dagger\|_u (1 + \gamma_2 \|x_0 - x^\dagger\|_u^\beta) + \delta, \tag{3.37}$$

hold for some $\gamma_2 > 0$ (independent of k).

We will now derive an estimate for δ in terms of k: (1.2), (2.2), $F(x^\dagger) = y$, (3.15), (3.23), and (3.29) imply that

$$\begin{aligned}
(\tau - 1)\delta &\leq \|y^\delta - F(x_j^\delta)\| - \|y^\delta - y\| \\
&\leq \|y - F(x_j^\delta)\| \leq \|q_j^\delta\| + \|e_j^\delta\|_{-a} \\
&\leq (1 + \tfrac{c}{\beta+1} \gamma^{a-b-u\beta} \|e_j^\delta\|_u^\beta) \|e_j^\delta\|_{-a}
\end{aligned} \tag{3.38}$$

for all $0 \leq j < k < k_*$ and hence (3.33) and (3.34) for $j = k-1$ yield that

$$\delta \leq \tfrac{\tau}{2(\tau-1)} \eta k^{-\frac{a+u}{2(a+s)}} \|x_0 - x^\dagger\|_u \tag{3.39}$$

provided that $c\gamma^{a-b-u\beta}\eta^\beta \|x_0 - x^\dagger\|_u^\beta/(\beta+1) \leq (\tau-2)/2$ which we assume in the following.

This together with (3.35) and (3.36) implies that

$$\|e_k^\delta\|_u \leq \|x_0 - x^\dagger\|_u (1 + \gamma_2 \|x_0 - x^\dagger\|_u^\beta + \tfrac{\tau}{2(\tau-1)}\eta) \leq \eta \|x_0 - x^\dagger\|_u$$

if $\gamma_2 \|x_0 - x^\dagger\|_u^\beta \le 1$ which we again assume to hold in the following. This proves (3.33) for $j = k$. Note that (3.38) is now also applicable to $j = k$. Together with (3.33) this implies that

$$\delta \le \tfrac{\tau}{2(\tau-1)} \|e_k^\delta\|_{-a}$$

and hence together with (3.35) and (3.37) that

$$\begin{aligned} \|e_k^\delta\|_{-a} &\le \tfrac{2(\tau-1)}{\tau-2}(k+1)^{-\frac{a+u}{2(a+s)}} \|x_0 - x^\dagger\|_u (1 + \gamma_2 \|x_0 - x^\dagger\|_u^\beta) \\ &\le \eta(k+1)^{-\frac{a+u}{2(a+s)}} \|x_0 - x^\dagger\|_u \end{aligned}$$

which proves (3.34) for $j = k$.

Since (3.36) and (3.39) are also valid for $k = k_*$, (3.33) also holds for $j = k_*$. Thus, if $\|x_0 - x^\dagger\|_u$ is sufficiently small, namely if

$$\|x_0 - x^\dagger\|_u \le \min\{\rho\gamma^u\eta^{-1},\ (\tau-2)(\beta+1)(2c)^{-1}\gamma^{u\beta+b-a}\eta^{-\beta},\ \gamma_2^{-1}\}, \tag{3.40}$$

then (3.12), (3.33), (3.34), and (3.35) imply the estimate (3.31).

Since similar to the derivation of (3.38) we obtain that

$$\begin{aligned} \|y^\delta - F(x_k^\delta)\| &\le \delta + (1 + \tfrac{c}{\beta+1}\gamma^{a-b-u\beta} \|e_k^\delta\|_u^\beta) \|e_k^\delta\|_{-a} \\ &\le \tfrac{\tau}{\tau-1}(1 + \tfrac{c}{\beta+1}\gamma^{a-b-u\beta} \|e_k^\delta\|_u^\beta) \|e_k^\delta\|_{-a}\,, \end{aligned}$$

the estimate (3.32) now follows with (3.33), (3.34), (3.35), and (3.40).

In the case of exact data ($\delta = 0$), the estimates hold for all $k \ge 0$, since then Lemma 3.6 holds for all $k \ge 0$. □

Theorem 3.8 *Under the assumptions of Proposition* 3.7 *the following estimates are valid for* $\delta > 0$ *and some positive constants* c_r*:*

$$k_* \le \left(\tfrac{2\tau}{\tau-2} \|x_0 - x^\dagger\|_u \delta^{-1}\right)^{\frac{2(a+s)}{a+u}} \tag{3.41}$$

and for $-a \le r < u$

$$\|x_{k_*}^\delta - x^\dagger\|_r \le c_r \|x_0 - x^\dagger\|_u^{\frac{a+r}{a+u}} \delta^{\frac{u-r}{a+u}}\,. \tag{3.42}$$

Proof. The estimate (3.41) follows immediately from (3.35) and (3.39) (with $k = k_*$).

We will now derive an estimate for $\|e_{k_*}^\delta\|_{-a}$. Combining $F(x^\dagger) = y$, (1.2), (2.2), (3.23), (3.29), and (3.33) (with $j = k_*$) we obtain that

$$\begin{aligned} \|e_{k_*}^\delta\|_{-a} &\le \|q_{k_*}^\delta\| + \delta + \|y^\delta - F(x_{k_*}^\delta)\| \\ &\le \tfrac{c}{\beta+1}\gamma^{a-b-u\beta}\eta^\beta \|x_0 - x^\dagger\|_u^\beta \|e_{k_*}^\delta\|_{-a} + (\tau+1)\delta \end{aligned}$$

and hence that

$$\|e_{k_*}^\delta\|_{-a} \le \tilde\gamma\delta$$

for some $\tilde\gamma > 0$ if $c\gamma^{a-b-u\beta}\eta^\beta \|x_0 - x^\dagger\|_u^\beta/(\beta+1) < 1$, which is a further restriction on $\|x_0 - x^\dagger\|_u$ in addition to (3.40) if $\tau \ge 4$.

Thus for $\|x_0 - x^\dagger\|_u$ sufficiently small, we now obtain with (3.12) and (3.33) (with $j = k_*$) that

$$\|e_{k_*}^\delta\|_r \le \|e_{k_*}^\delta\|_{-a}^{\frac{u-r}{a+u}} \|e_{k_*}^\delta\|_u^{\frac{a+r}{a+u}} \le (\tilde\gamma\delta)^{\frac{u-r}{a+u}} (\eta\|x_0 - x^\dagger\|_u)^{\frac{a+r}{a+u}}$$

which verifies (3.42). □

As usual for regularization in Hilbert scales, we are interested in obtaining convergence rates with respect to the norm in $\mathcal{X} = \mathcal{X}_0$.

Corollary 3.9 *Under the assumptions of Proposition* 3.7 *the following estimates hold:*

$$\|x_{k_*}^\delta - x^\dagger\|_0 = O\Big(\delta^{\frac{u}{a+u}}\Big) \qquad \text{if} \quad s \le 0, \tag{3.43}$$

$$\|x_{k_*}^\delta - x^\dagger\|_0 = O\Big(\|x_{k_*}^\delta - x^\dagger\|_s\Big) = O\Big(\delta^{\frac{u-s}{a+u}}\Big) \qquad \text{if} \quad 0 < s < u. \tag{3.44}$$

If in addition (3.3) *holds, then for* $s > 0$ *the rate can be improved to*

$$\|x_{k_*}^\delta - x^\dagger\|_0 = O\Big(\|x_{k_*}^\delta - x^\dagger\|_r\Big) = O\Big(\delta^{\frac{u-r}{a+u}}\Big) \quad \text{if} \quad r := \tfrac{s(\tilde a - a)}{\tilde a + s} \le u. \tag{3.45}$$

Proof. The proof follows immediately from Proposition 3.3 (iii), (iv) and Theorem 3.8. □

Remark 3.10 Note that (3.41) implies that k_* is finite for $\delta > 0$ and hence $x_{k_*}^\delta$ is a stable approximation of $x^\dagger$.

Moreover, it can be seen from (3.41) that the larger s the faster k_* possibly grows if $\delta \to 0$. As a consequence, s should be kept as small as possible to reduce the number of iterations and hence to reduce the numerical effort: this means that, in the case $b = a$, $s > 0$ is only necessary if $u > a$. However, if F' is only locally Lipschitz continuous in $\widetilde{\mathcal{X}}_0$ (i.e., $b = 0$, $\beta = 1$), then $a < u \le 2s$ and hence $s > a/2$ is necessary.

On the other hand, if u is close to 0, it might be possible to choose a negative s. According to (3.43), we would still get the optimal rate, but, due to (3.41), k_* would not grow so fast. Choosing a negative s could be interpreted as a preconditioned Landweber method (cf. [42]). If for instance $\beta = 1$, $a/2 < b \le a$, and $a - b < u < b$, then a choice $s = (u - b)/2 < 0$ is possible (cf. Assumption 3.4 (iii)).

We will now comment on the rates in Corollary 3.9: if only Assumption 3.1 (iii) is satisfied, i.e., if $\|F'(x^\dagger)x\|$ may be estimated through the norm in $\mathcal{X}_{-a}$ only from above, convergence rates in $\mathcal{X}_0$ can only be given if $s < u$, i.e., only for the case of undersmoothing. If $s > 0$ the rates will not be optimal in general. To obtain rates also for $s > u$, i.e., for the case of oversmoothing, condition (3.3) has to be additionally satisfied. From what we said on the choice of s above, the case of oversmoothing is not desirable. However, note that the rates for $\|x_{k_*}^\delta - x^\dagger\|_0$ can be improved if (3.3) holds also for $0 < s < u$. Moreover, if $\tilde{a} = a$, i.e., if the usual equivalence condition (3.2) is satisfied, then we always obtain the usual optimal rates $O(\delta^{\frac{u}{a+u}})$ (see [123]).

For numerical computations one has to approximate the infinite-dimensional spaces by finte-dimensional ones. Also the operators F and $F'(x)^*$ have to be approximated by suitable finite-dimensional realizations. An appropriate convergence rates analysis has been carried out in [124]. This analysis also shows that a modification, where $F'(x_k^\delta)^*$ in (3.1) is replaced by $G^\delta(x_k^\delta)$ similar as in (2.24), is possible.

Numerical results, where Landweber iteration in Hilbert scales has been applied to a nonlinear Hammerstein equation, can be found in [124]. The results there show that the correct choice of Hilbert scale can reduce the number of iterations to obtain the same quality as for standard Landweber iteration tremendously.

3.2 Iteratively regularized Landweber iteration

In [144], the following modification of Landweber iteration was suggested:

$$x_{k+1}^\delta = x_k^\delta + F'(x_k^\delta)^*(y^\delta - F(x_k^\delta)) + \beta_k(x_0 - x_k^\delta) \qquad \text{with} \quad 0 < \beta_k \le \beta_{\max} < \tfrac{1}{2}\,. \tag{3.46}$$

The additional term $\beta_k(x_k^\delta - x_0)$ compared to the classical Landweber iteration (2.1) is motivated by the iteratively regularized Gauss–Newton method, see Section 4.2.

First we show that under certain conditions x_k^δ remains in a ball around x_0.

Proposition 3.11 *Assume that $F(x) = y$ has a solution $x_* \in \mathcal{B}_\rho(x_0)$ and let $\kappa \in (0,1)$ be a fixed positive constant and*

$$c(\rho) := \rho\, \frac{1 - \beta_{\max} + \sqrt{1 + \beta_{\max}(2 - \beta_{\max})\kappa^{-2}}}{2 - \beta_{\max}}\,. \tag{3.47}$$

(i) *Let the conditions* (2.3) *and* (2.4) *hold on* $\mathcal{B}_{\rho+c(\rho)}(x_0)$ *with* η *in* (2.4) *satisfying*

$$E := 1 - 2\eta - 2\beta_{\max}(1-\eta) - \kappa^2 > 0\,. \tag{3.48}$$

If $x_k^\delta \in \mathcal{B}_{c(\rho)}(x_*)$, *then a sufficient condition for* $x_{k+1}^\delta \in \mathcal{B}_{c(\rho)}(x_*) \subset \mathcal{B}_{\rho+c(\rho)}(x_0)$ *is that*

$$\|y^\delta - F(x_k^\delta)\| > 2(1-\beta_{\max})\frac{1+\eta}{E}\,\delta\,. \tag{3.49}$$

(ii) *Assume that the conditions* (2.3) *and* (2.4) *hold on* $\mathcal{B}_{\rho+\hat{c}(\rho)}(x_0)$ *with* η *in* (2.4) *satisfying* (3.48) *and*

$$\hat{c}(\rho) := c(\rho) + D(1+\eta) + \tfrac{1}{2}(D\beta_{\max} + \rho) \tag{3.50}$$

for some $D > 0$. *If* $x_k^\delta \in \mathcal{B}_{\hat{c}(\rho)}(x_*)$, *a sufficient condition for* $x_{k+1}^\delta \in \mathcal{B}_{\hat{c}(\rho)}(x_*) \subset \mathcal{B}_{\rho+\hat{c}(\rho)}(x_0)$ *is that*

$$\frac{\delta}{\beta_k} \le D\,. \tag{3.51}$$

Proof. It follows immediately from the definition (3.46) that

$$\begin{aligned}
\|x_{k+1}^\delta - x_*\|^2 \;=\;& (1-\beta_k)^2\,\|x_k^\delta - x_*\|^2 + \beta_k^2\,\|x_0 - x_*\|^2 \\
&+ 2\beta_k(1-\beta_k)\langle\, x_k^\delta - x_*, x_0 - x_*\,\rangle \\
&+ \|F'(x_k^\delta)^*(y^\delta - F(x_k^\delta))\|^2 \\
&+ 2(1-\beta_k)\langle\, y^\delta - F(x_k^\delta), F'(x_k^\delta)(x_k^\delta - x_*)\,\rangle \\
&+ 2\beta_k\langle\, x_0 - x_*, F'(x_k^\delta)^*(y^\delta - F(x_k^\delta))\,\rangle\,.
\end{aligned} \tag{3.52}$$

This estimate together with (1.2), (2.3), (2.4), and the formulae

$$\begin{aligned}
&\langle\, y^\delta - F(x_k^\delta), F'(x_k^\delta)(x_k^\delta - x_*)\,\rangle \\
&\quad= \langle\, y^\delta - F(x_k^\delta), y^\delta - y\,\rangle - \|y^\delta - F(x_k^\delta)\|^2 \\
&\qquad - \langle\, y^\delta - F(x_k^\delta), F(x_k^\delta) - F(x_*) - F'(x_k^\delta)(x_k^\delta - x_*)\,\rangle
\end{aligned} \tag{3.53}$$

and

$$2\beta_k\,\|x_0 - x_*\|\,\|y^\delta - F(x_k^\delta)\| \le \kappa^{-2}\beta_k^2\,\|x_0 - x_*\|^2 + \kappa^2\,\|y^\delta - F(x_k^\delta)\|^2$$

yields that

$$\begin{aligned} \|x_{k+1}^\delta - x_*\|^2 &\leq (1-\beta_k)^2 \|x_k^\delta - x_*\|^2 + \beta_k^2(1+\kappa^{-2}) \|x_0 - x_*\|^2 \\ &\quad + 2\beta_k(1-\beta_k) \|x_k^\delta - x_*\| \, \|x_0 - x_*\| \\ &\quad + \|y^\delta - F(x_k^\delta)\| \Big(2(1-\beta_k)(1+\eta)\delta \\ &\quad - \|y^\delta - F(x_k^\delta)\| (1 - 2\eta - 2\beta_k(1-\eta) - \kappa^2) \Big) . \end{aligned} \tag{3.54}$$

We first consider case (i). For $0 \leq \beta_k \leq \beta_{\max}$, (3.48) implies that

$$\frac{1-\beta_{\max}}{1 - 2\eta - 2\beta_{\max}(1-\eta) - \kappa^2} \geq \frac{1-\beta_k}{1 - 2\eta - 2\beta_k(1-\eta) - \kappa^2} .$$

Hence, we obtain together with (3.49), (3.54), $\|x_0 - x_*\| \leq \rho$, and $\|x_k^\delta - x_*\| \leq c(\rho)$ that

$$\|x_{k+1}^\delta - x_*\|^2 \leq (1-\beta_k)^2 c(\rho)^2 + \beta_k^2(1+\kappa^{-2})\rho^2 + 2\beta_k(1-\beta_k)\rho c(\rho) \leq c(\rho)^2 .$$

The last inequality follows from (3.47) and the fact that

$$(1-\beta_k)^2 c^2 + \beta_k^2(1+\kappa^{-2})\rho^2 + 2\beta_k(1-\beta_k)\rho c \leq c^2$$

if

$$c \geq \rho \frac{1 - \beta_k + \sqrt{1 + \beta_k(2-\beta_k)\kappa^{-2}}}{2 - \beta_k} ,$$

which is true for $c(\rho)$.

Now we turn to case (ii). From (1.2) and (2.3) it follows that

$$\begin{aligned} \|y^\delta - F(x_k^\delta)\| &\leq \|y^\delta - y\| + \int_0^1 \|F'(x_* + t(x_k^\delta - x_*))(x_k^\delta - x_*)\| \, dt \\ &\leq \delta + \|x_k^\delta - x_*\| . \end{aligned}$$

This together with $\beta_k \leq \beta_{\max}$, (3.48), (3.51), (3.54), and $\|x_0 - x_*\| \leq \rho$ yields the estimate

$$\begin{aligned} &\|x_{k+1}^\delta - x_*\|^2 + E\|y^\delta - F(x_k^\delta)\|^2 \\ &\quad\leq (1-\beta_k)^2 \|x_k^\delta - x_*\|^2 \\ &\qquad + \beta_k^2 \Big((1+\kappa^{-2})\rho^2 + 2D^2(1-\beta_k)(1+\eta) \Big) \\ &\qquad + 2\beta_k(1-\beta_k) \|x_k^\delta - x_*\| (\rho + D(1-\beta_k)) \end{aligned} \tag{3.55}$$

and furthermore with $\|x_k^\delta - x_*\| \le \hat{c}(\rho)$ that

$$\begin{aligned}\|x_{k+1}^\delta - x_*\|^2 &\le (1-\beta_k)^2 \hat{c}(\rho)^2 + \beta_k^2\Big((1+\kappa^{-2})\rho^2 + 2D^2(1-\beta_k)(1+\eta)\Big) \\ &\quad + 2\beta_k(1-\beta_k)\hat{c}(\rho)(\rho + D(1+\eta)) \le \hat{c}(\rho)^2 .\end{aligned}$$

The correctness of the last inequality follows similar as for case (i) noting that

$$\begin{aligned}&\frac{(1-\beta_k)a + \sqrt{(1-\beta_k)^2a^2 + \beta_k(2-\beta_k)(p+2(1-\beta_k)q)}}{2-\beta_k} \\ &\quad\le c(\rho) + \tfrac{1}{2}D(1+\eta) + \tfrac{1}{2}\sqrt{D(1+\eta)(D(1+\eta+2\beta_{\max}) + 2\rho)} \\ &\quad\le c(\rho) + D(1+\eta) + \tfrac{1}{2}(D\beta_{\max} + \rho) = \hat{c}(\rho) ,\end{aligned}$$

where $a := \rho + D(1+\eta)$, $p := (1+\kappa^{-2})\rho^2$, and $q := D^2(1+\eta)$. □

Note that if $\beta_{\max} = 0$, then in case (i) κ may be chosen as 0 and hence (3.49) reduces to condition (2.12), which is required to prove monotonicity of $\|e_k^\delta\|$ for the classical Landweber iteration (2.1).

Since $\hat{c}(\rho) \ge c(\rho)$, in case (ii) the conditions (2.3) and (2.4) have to be satisfied on a larger domain than in case (i).

The proposition above gives rise to the following two stopping criteria for the iteratively regularized Landweber iteration:

The first is an a-posteriori stopping rule, namely the discrepancy principle (2.2), i.e., the iteration is terminated after $k_* = k_*(\delta, y^\delta)$ steps with

$$\|y^\delta - F(x_{k_*}^\delta)\| \le \tau\delta < \|y^\delta - F(x_k^\delta)\| , \qquad 0 \le k < k_* ,$$

where

$$\tau > 2(1-\beta_{\max})\frac{1+\eta}{E} \tag{3.56}$$

and E is as in (3.48).

The second is an a-priori stopping rule, where the iteration is terminated after $k_* = k_*(\delta)$ steps with

$$D\beta_{k_*} < \delta \le D\beta_k , \qquad 0 \le k < k_* , \tag{3.57}$$

for some $D > 0$. If $\beta_k \to 0$ as $k \to \infty$, then obviously in (3.57) $k_*(\delta) < \infty$ for $\delta > 0$.

Next we prove a result similar to Corollary 2.3 for the classical Landweber iteration.

Corollary 3.12 *Assume that $F(x) = y$ has a solution $x_* \in \mathcal{B}_\rho(x_0)$.*

(i) Let the assumptions of Proposition 3.11 (i) hold. If the iteratively regularized Landweber iteration is stopped according to the a-posteriori stopping rule (2.2), (3.56), *then*

$$k_*(\tau\delta)^2 < \sum_{k=0}^{k_*-1} \|y^\delta - F(x_k^\delta)\|^2 \le \Psi^{-1}\rho^2\Big(1 + 2(1+\kappa^{-2})\sum_{k=0}^{k_*-1}\beta_k\Big)$$

with

$$\Psi := E - 2\tau^{-1}(1-\beta_{\max})(1+\eta) > 0\,. \tag{3.58}$$

In particular, if $y^\delta = y$ (i.e., if $\delta = 0$), then

$$\sum_{k=0}^{\infty} \|y - F(x_k)\|^2 \le E^{-1}\rho^2\Big(1 + 2(1+\kappa^{-2})\sum_{k=0}^{\infty}\beta_k\Big).$$

(ii) Let the assumptions of Proposition 3.11 (ii) hold. If the iteratively regularized Landweber iteration is stopped according to the a-priori stopping rule (3.57), *then it follows that*

$$\sum_{k=0}^{k_*-1} \|y^\delta - F(x_k^\delta)\|^2 \le E^{-1}\Big(\rho^2 + \Phi\sum_{k=0}^{k_*-1}\beta_k\Big)$$

with

$$\Phi := 2(\rho + D(1+\eta))(\rho(1+\kappa^{-2}) + D(1+\eta) + \tfrac{1}{2}(D\beta_{\max} + \rho)\,.$$

Proof. We first consider case (i). Since $x_0^\delta = x_0 \in \mathcal{B}_\rho(x_*)$, it follows by induction that Proposition 3.11 (i) is applicable for all $0 \le k < k_*$. Therefore, it follows with (3.47) that $\|x_k^\delta - x_*\| \le c(\rho) < \rho\sqrt{1+\kappa^{-2}} < \rho(1+\kappa^{-2})$. Moreover, we obtain with (3.54) and (3.58) that

$$\begin{aligned} &\|x_{k+1}^\delta - x_*\|^2 + \Psi\,\|y^\delta - F(x_k^\delta)\|^2 \\ &\quad\le\; (1-\beta_k)^2\,\|x_k^\delta - x_*\|^2 + \beta_k^2(1+\kappa^{-2})\rho^2 \\ &\qquad + 2\beta_k(1-\beta_k)\,\|x_k^\delta - x_*\|\,\rho \\ &\quad\le\; \|x_k^\delta - x_*\|^2 + 2\rho^2(1+\kappa^{-2})\beta_k\,. \end{aligned} \tag{3.59}$$

Note that, due to (3.48) and (3.56), $\Psi > 0$. Thus, the assertion of case (i) now follows by induction.

In the noise free case, k_* may be any positive integer and τ may be chosen arbitrarily large yielding the appropriate estimate for $\delta = 0$. Note that Ψ may be replaced by E then.

Let us now turn to case (ii). It follows immediately from Proposition 3.11 (ii) and (3.55) that

$$\begin{aligned} &\|x_{k+1}^\delta - x_*\|^2 + E\,\|y^\delta - F(x_k^\delta)\|^2 \\ &\quad\le \|x_k^\delta - x_*\|^2 + \beta_k^2\Big((1+\kappa^{-2})\rho^2 + 2D^2(1-\beta_k)(1+\eta)\Big) \\ &\quad\quad + 2\beta_k(1-\beta_k)\hat{c}(\rho)(\rho + D(1+\eta)) \\ &\quad\le \|x_k^\delta - x_*\|^2 + \Phi\beta_k\,, \end{aligned} \tag{3.60}$$

where we used the fact that $\hat{c}(\rho) \le \rho(1+\kappa^{-2}) + D(1+\eta) + \frac{1}{2}(D\beta_{\max} + \rho)$. The assertion of case (ii) now follows as above by induction. □

It follows from the corollary above that the termination index $k_*(\delta, y^\delta)$ determined by the discrepancy principle (2.2), (3.56) is finite for $\delta > 0$ if the following condition

$$\sum_{k=0}^{\infty} \beta_k < \infty \tag{3.61}$$

holds.

The same condition is needed to prove convergence of the iteratively regularized Landweber iteration for the noise free case.

Theorem 3.13 *Assume that the conditions* (2.3) *and* (2.4) *hold on* $\mathcal{B}_{\rho+c(\rho)}(x_0)$ *with* $c(\rho)$ *as in* (3.47) *and* η *satisfying* (3.48). *Moreover, assume that* $F(x) = y$ *is solvable in* $\mathcal{B}_\rho(x_0)$ *and that* $\{\beta_k\}$ *satisfies* (3.61). *Then the iteratively regularized Landweber iteration applied to exact data* y *converges to a solution of* $F(x) = y$ *in* $\mathcal{B}_{\rho+c(\rho)}(x_0)$. *If* $\mathcal{N}(F'(x^\dagger)) \subset \mathcal{N}(F'(x))$ *for all* $x \in \mathcal{B}_{\rho+c(\rho)}(x_0)$, *where* $x^\dagger$ *denotes the unique* x_0*-minimum-norm solution, then* x_k *converges to* $x^\dagger$ *as* $k \to \infty$.

Proof. The proof is similar to the one of Theorem 2.4 for Landweber iteration. However, now we cannot use the monotonicity of the solution errors $\|e_k\|$, where $e_k := x_k - x^\dagger$.

From Proposition 3.11 (i) it follows that the sequence $\{\,\|e_k\|\,\}$ is bounded, and hence it has a convergent subsequence $\{\,\|e_{k_n}\|\,\}$ with limit $\varepsilon \ge 0$. It follows immediately from (3.59) that

$$\|e_{k+1}\| \le (1-\beta_k)\,\|e_k\| + \beta_k\rho\sqrt{1+\kappa^{-2}}$$

and therefore by induction that

$$\|e_k\| \le \|e_m\| \prod_{l=m}^{k-1}(1-\beta_l) + \rho\sqrt{1+\kappa^{-2}}\Big(1-\prod_{l=m}^{k-1}(1-\beta_l)\Big) \tag{3.62}$$

for $m < k$, where we have used the formula

$$1-\prod_{l=m}^{n}(1-\beta_l) = \sum_{j=m}^{n}\beta_j \prod_{l=j+1}^{n}(1-\beta_l)\,, \tag{3.63}$$

which is satisfied for $n \ge m$ with the usual convention that $\prod_{l=n+1}^{n}(1-\beta_l) = 1$.
If $k_{n-1} < k < k_n$, then it follows from (3.62) that

$$\begin{aligned}
\|e_k\| &\le \|e_{k_{n-1}}\| + (\rho\sqrt{1+\kappa^{-2}} - \|e_{k_{n-1}}\|)\Big(1-\prod_{l=k_{n-1}}^{k-1}(1-\beta_l)\Big),\\
\|e_k\| &\ge \|e_{k_n}\| - (\rho\sqrt{1+\kappa^{-2}} - \|e_k\|)\Big(1-\prod_{l=k}^{k_n-1}(1-\beta_l)\Big).
\end{aligned}$$

These estimates together with the convergence of $\|e_{k_n}\|$ immediately yield that

$$\lim_{k\to\infty} \|e_k\| = \varepsilon\,,$$

since (3.61) implies that

$$\prod_{l=m}^{k-1}(1-\beta_l) \to 1 \quad \text{as} \quad m\to\infty\,, \qquad k > m\,. \tag{3.64}$$

Note that $k_{n-1}, k_n \to \infty$ as $k \to \infty$.

Now we proceed as in Theorem 2.4: if for $n \ge k$ we choose m with $n \ge m \ge k$ such that

$$\|y - F(x_m)\| \le \|y - F(x_i)\| \qquad \text{for all} \qquad k \le i \le n\,,$$

then convergence of the iterates x_k is guaranteed, if we can show that the terms $\langle\, e_m - e_n, e_m\,\rangle$ and $\langle\, e_m - e_k, e_m\,\rangle$ tend to zero as $k \to \infty$. Since, due to Corollary 3.12 (i) and (3.61), the residuals $y - F(x_k)$ converge to zero as $k \to \infty$, the limit of x_k must then be a solution of $F(x) = y$.

From (3.46) and (3.63) we obtain by induction the representation

$$\begin{aligned}
e_k = {} & e_m \prod_{l=m}^{k-1}(1-\beta_l) + \sum_{j=m}^{k-1} F'(x_j)^*(y-F(x_j)) \prod_{l=j+1}^{k-1}(1-\beta_l)\\
&+ (x_0 - x^\dagger)\Big(1-\prod_{l=m}^{k-1}(1-\beta_l)\Big)
\end{aligned}$$

for $m < k$. Using this formula, we obtain the estimate

$$\begin{aligned} |\langle e_m - e_n, e_m \rangle| &\leq \Big(1 - \prod_{l=m}^{n-1}(1-\beta_l)\Big)|\langle x_0 - x_m, e_m \rangle| \\ &\quad + \sum_{j=m}^{n-1}\Big(\prod_{l=j+1}^{n-1}(1-\beta_l)\Big)|\langle y - F(x_j), F'(x_j)(x_m - x^\dagger)\rangle|. \end{aligned}$$

Since $|\langle x_0 - x_m, e_m \rangle|$ is bounded, the first term on the right hand side tends to zero due to (3.64). Since $(1-\beta_l) \leq 1$, the second term on the right hand side may be estimated by

$$(1+3\eta)\sum_{j=m}^{n-1}\|y - F(x_j)\|^2$$

as in Theorem 2.4. Due to (3.61), it follows from Proposition 3.11 (i) that this sum tends to zero as $m \to \infty$. This shows that $\langle e_m - e_n, e_m \rangle \to 0$. Analogously one verifies that $\langle e_m - e_k, e_m \rangle \to 0$.

The last assertion on the x_0-minimum-norm solution follows as in the proof of Theorem 2.4. □

The next result shows that the iteratively regularized Landweber iteration together with the a-posteriori stopping rule (2.2), (3.56) is a regularization method.

Theorem 3.14 *Let the assumptions of Theorem* 3.13 *hold and let* $k_* = k_*(\delta, y^\delta)$ *be chosen according to the stopping rule* (2.2), (3.56). *Then the iteratively regularized Landweber iterates* $x^\delta_{k_*}$ *converge to a solution of* $F(x) = y$. *If* $\mathcal{N}(F'(x^\dagger)) \subset \mathcal{N}(F'(x))$ *for all* $x \in \mathcal{B}_{\rho+c(\rho)}(x_0)$, *then* $x^\delta_{k_*}$ *converges to* $x^\dagger$ *as* $\delta \to 0$.

Proof. The proof is analogous to the one of Theorem 2.6. □

In the next theorem we verify the regularization properties of the iteratively regularized Landweber iteration if it is terminated by the a-priori rule (3.57).

Theorem 3.15 *Assume that the conditions* (2.3) *and* (2.4) *hold on* $\mathcal{B}_{\rho+\hat{c}(\rho)}(x_0)$ *with* $\hat{c}(\rho)$ *as in* (3.50) *and* η *satisfying* (3.48). *Moreover, assume that* $F(x) = y$ *is solvable in* $\mathcal{B}_\rho(x_0)$ *and that* $\{\beta_k\}$ *satisfies* (3.61). *In addition, let* $k_* = k_*(\delta)$ *be chosen according to the stopping rule* (3.57). *Then the iteratively regularized Landweber iterates* $x^\delta_{k_*}$ *converge to a solution of* $F(x) = y$. *If* $\mathcal{N}(F'(x^\dagger)) \subset \mathcal{N}(F'(x))$ *for all* $x \in \mathcal{B}_{\rho+\hat{c}(\rho)}(x_0)$, *then* $x^\delta_{k_*}$ *converges to* $x^\dagger$ *as* $\delta \to 0$.

Proof. Since $\hat{c}(\rho) \geq c(\rho)$, Theorem 3.13 is applicable. Let x_* be the limit of the iteratively regularized Landweber iteration with precise data y and let $\{\delta_n\}$ be a sequence converging to zero as $n \to \infty$. Denote by $y_n := y^{\delta_n}$ a corresponding sequence of perturbed data, and by $k_n = k_*(\delta_n)$ the stopping index determined by (3.57). Since $\beta_k > 0$ and since, due to (3.61), $\beta_k \to 0$ as $k \to \infty$, it holds that $k_n \to \infty$. Then for $k_n > k$ we conclude from (3.60) that

$$\|x_{k_n}^{\delta_n} - x_*\|^2 \leq (\|x_k^{\delta_n} - x_k\| + \|x_k - x_*\|)^2 + \Phi \sum_{l=k}^{k_n-1} \beta_k \,.$$

Using (3.61), the rest of the proof is analogous to the one of Theorem 2.6. □

Convergence rates results for the classical Landweber iteration were obtained only under restrictive conditions on the nonlinearity of F (see Assumption 2.7). We will show in the following theorem that similarly to Tikhonov regularization convergence rates for iteratively regularized Landweber iteration may be obtained under less restrictive assumptions on F if a sufficiently strong source condition is satisfied. Even condition (2.4) is not needed.

Assumption 3.16 Let ρ be a positive number such that $\mathcal{B}_{2\rho}(x_0) \subset \mathcal{D}(F)$.

(i) The equation $F(x) = y$ has an x_0-minimum-norm solution $x^\dagger$ in $\mathcal{B}_\rho(x_0)$ satisfying the source condition

$$x^\dagger - x_0 = F'(x^\dagger)^* w \,, \quad w \in \mathcal{N}(F'(x^\dagger)^*)^\perp .$$

This corresponds to (2.22) with $\mu = 1/2$ (cf. [45, Proposition 2.18]).

(ii) The operator F is Fréchet-differentiable and satisfies $\|F'(x)\| \leq \omega \leq 1$ in $\mathcal{B}_{2\rho}(x_0)$. If $\omega < 1$, then this condition is stronger than the previously used scaling condition (2.3).

(iii) The Fréchet-derivative is locally Lipschitz continuous, i.e.,

$$\|F'(x) - F'(\tilde{x})\| \leq L\|x - \tilde{x}\| \,, \qquad x, \tilde{x} \in \mathcal{B}_{2\rho}(x_0) \,.$$

Theorem 3.17 *Let Assumption 2.7 hold, let $\{\beta_k\}$ be monotonically decreasing, and assume that positive parameters a, b, c, d exist such that the following conditions hold:*

$$(1 + a)\omega^2 + b + c + d + 2\beta_0 - 2 \leq 0 \,, \tag{3.65}$$

$$\|x_0 - x^\dagger\|^2 \leq 2\beta_0 \sqrt{qdL^{-1}} \leq \rho^2 \,, \tag{3.66}$$

$$\sqrt{qd^{-1}L} < p \leq \beta_0^{-1} \,, \tag{3.67}$$

$$\beta_k(1-\beta_k(p-\sqrt{qd^{-1}L}))\le\beta_{k+1}\,, \tag{3.68}$$

where

$$q:=\|w\|^2(\omega^2(1+a^{-1})+b^{-1})+2\|w\|\,D+c^{-1}D^2\,, \tag{3.69}$$

$$p:=1+(1-L\|w\|)(1-\beta_0))\,, \tag{3.70}$$

and D is as in (3.57). *If the iteratively regularized Landweber iteration is terminated with the a-priori stopping criterion* (3.57), *then it holds that*

$$\|x^\delta_{k_*}-x^\dagger\|=O(\sqrt{\delta})\,.$$

In the noise free case ($\delta=D=0$), we get the rate

$$\|x_k-x^\dagger\|=O(\sqrt{\beta_k})\,.$$

Proof. Let us assume that $x^\delta_k\in\mathcal{B}_\rho(x^\dagger)\subset\mathcal{B}_{2\rho}(x_0)$ for some $0\le k<k_*$ and note that, due to the Lipschitz continuity of F', it holds that

$$\|F(x)-F(\tilde{x})-F'(\tilde{x})(x-\tilde{x})\|\le\tfrac{1}{2}L\,\|x-\tilde{x}\|^2\,,\quad x,\tilde{x}\in\mathcal{B}_{2\rho}(x_0)\,. \tag{3.71}$$

Together with (1.2), (3.52), (3.53), and Assumption 3.16, we then obtain the estimate

$$\begin{aligned}\|x^\delta_{k+1}-x^\dagger\|^2 \le\; & (1-\beta_k)^2\,\|x^\delta_k-x^\dagger\|^2+\beta_k^2\omega^2\,\|w\|^2\\ & +2\beta_k(1-\beta_k)\,\|w\|\,(\delta+\|y^\delta-F(x^\delta_k)\|+\tfrac{1}{2}L\,\|x^\delta_k-x^\dagger\|^2)\\ & +\omega^2\,\|y^\delta-F(x^\delta_k)\|^2+2\beta_k\omega^2\,\|w\|\,\|y^\delta-F(x^\delta_k)\|\\ & +2(1-\beta_k)\,\|y^\delta-F(x^\delta_k)\|\,(\delta-\|y^\delta-F(x^\delta_k)\|\\ & \qquad\qquad +\tfrac{1}{2}L\,\|x^\delta_k-x^\dagger\|^2)\\ \le\; & \|x^\delta_k-x^\dagger\|^2(1-\beta_k)(1-\beta_k(1-L\,\|w\|))\\ & +\beta_k^2\omega^2\,\|w\|^2(1+a^{-1})+2\beta_k\,\|w\|\,\delta+b^{-1}\beta_k^2\,\|w\|^2+c^{-1}\delta^2\\ & +\|y^\delta-F(x^\delta_k)\|^2((1+a)\omega^2+b+c+d+2\beta_0-2)\\ & +\tfrac{1}{4}d^{-1}L\,\|x^\delta_k-x^\dagger\|^4\,.\end{aligned}$$

Note that $2AB\le s^{-1}A^2+sB^2$ for all $s,A,B>0$. Using (3.57), (3.65), (3.69), (3.70), and setting

$$\gamma^\delta_k:=\beta_k^{-1}\,\|x^\delta_k-x^\dagger\|^2$$

then yields that

$$\gamma^\delta_{k+1}\le\beta_{k+1}^{-1}\beta_k\Big(\gamma^\delta_k(1-p\beta_k)+q\beta_k+\tfrac{1}{4}d^{-1}L\beta_k(\gamma^\delta_k)^2\Big)\,. \tag{3.72}$$

We will now show by induction that x_k^δ remains in $\mathcal{B}_\rho(x^\dagger)$ and that $\gamma_k^\delta \leq 2\sqrt{qdL^{-1}}$ for all $0 \leq k \leq k_*$. The assertion holds for $k = 0$ due to (3.66). Let us now assume that it holds for $k < k_*$. Using (3.67), (3.68), and (3.72) we now obtain that

$$\gamma_{k+1}^\delta \leq 2\sqrt{qdL^{-1}}\beta_{k+1}^{-1}\beta_k(1 - \beta_k(p - \sqrt{qd^{-1}L})) \leq 2\sqrt{qdL^{-1}}\,.$$

Moreover, (3.66) and the monotonicity of $\{\beta_k\}$ yield that $x_{k+1}^\delta \in \mathcal{B}_\rho(x^\dagger)$. This already proves the rate result for the noise free case, since $k_*(0) = \infty$. The rate for the noisy case follows from (3.57), since then $\beta_{k_*} \leq D^{-1}\delta$. □

Remark 3.18 We discuss the assumptions made in Theorem 3.17:

Since $\omega \leq 1$ and since $\beta_0 < \frac{1}{2}$, positive parameters a, b, c, d always exist satisfying conditions (3.65) – (3.67) as long as $\|w\|$ and D in (3.57) are sufficiently small.

To explain (3.68), we assume for the sake of simplicity that $p - \sqrt{qd^{-1}L} = 1$. If $\tilde{\beta}_k$ is defined by

$$\tilde{\beta}_{k+1} := \tilde{\beta}_k(1 - \tilde{\beta}_k)\,, \qquad \tilde{\beta}_0 < \tfrac{1}{2}\,,$$

then it holds that $\tilde{\beta}_k \searrow 0$ and $\tilde{\beta}_k \sim (k+1)^{-1}$. Moreover, for any other monotonically decreasing sequence $\{\beta_k\}$ satisfying

$$\beta_{k+1} \geq \beta_k(1 - \beta_k)\,, \qquad \beta_0 = \tilde{\beta}_0\,,$$

it holds that $\beta_k \geq \tilde{\beta}_k$. A possible choice is for instance given by

$$\beta_k = (k+3)^{-1}\,.$$

Unfortunately, for sequences $\{\beta_k\}$ satisfying (3.68) it always holds that $\sum_{k=0}^\infty \beta_k$ is divergent. Thus, these sequences are not compatible with the general convergence results of the iteratively regularized Landweber iteration above (cf. (3.61)).

Remark 3.19 In the proof of Theorem 3.17 it was nowhere used that $x^\dagger$ has to be an x_0-minimum-norm solution. Mereley a solution in $\mathcal{B}_\rho(x_0)$ is needed that satisfies the source conditions in Assumption 3.16 (i). However, if $x^\dagger - x_0 = F'(x^\dagger)^* w$ with $L\|w\| < 1$, then $x^\dagger \in \mathcal{B}_\rho(x_0)$ is already the unique x_0-minimum-norm solution as the following argument shows: let $x \in \mathcal{B}_\rho(x_0)$, $x \neq x^\dagger$, be such that $F(x) = y$. Then (3.71) implies that

$$\begin{aligned}
\|x - x_0\|^2 &= \|x - x^\dagger\|^2 + \|x^\dagger - x_0\|^2 + 2\langle F'(x^\dagger)(x - x^\dagger), w\rangle \\
&\geq \|x - x^\dagger\|^2 + \|x^\dagger - x_0\|^2 - 2\|w\|\,\|F'(x^\dagger)(x - x^\dagger)\| \\
&\geq \|x - x^\dagger\|^2(1 - L\|w\|) + \|x_0 - x^\dagger\|^2 > \|x_0 - x^\dagger\|^2.
\end{aligned}$$

Finally, we want to mention that an inspection of the proof of Theorem 3.17 shows that a convergence rate result can also be proven for iterates x_k^δ, where k_* is chosen according to the discrepancy principle (2.2), (3.56). Under similar conditions as in Theorem 3.17 one obtains the rate

$$\|x_{k_*}^\delta - x^\dagger\| = O(\sqrt{\beta_{k_*}})\,.$$

3.3 A derivative free approach

Based on an idea by Engl and Zou [47], Kügler, in his thesis [108] (see also [107]), developed a modification of Landweber iteration for parameter identification problems where it is not needed that F is Fréchet-differentiable.

In this approach, the nonlinear operator F is defined via an equation, i.e., $F(x) = y$ is the solution of the equation

$$C(x)y = f \tag{3.73}$$

where $C(x) : \mathcal{Y}_0 \to \mathcal{Y}_0^*$ and $f \in \mathcal{Y}_0^*$ is a known right hand side. $\mathcal{Y}_0$ is a closed subspace of the Hilbert space $\mathcal{Y}$ and $\mathcal{Y}_0^*$ is the dual space of $\mathcal{Y}_0$.

The following assumption is essential for the analysis developed in [108]: there exist positive constants α_1 and α_2 such that

$$\alpha_1 \|v - w\|^2 \le (C(x)v - C(x)w, v - w)\,, \qquad v, w \in \mathcal{Y}_0\,,$$

and

$$(C(x)v - C(x)w, y) \le \alpha_2 \|v - w\|\,\|y\|\,, \qquad v, w, y \in \mathcal{Y}$$

hold for all $x \in \mathcal{D}(F) \subset \mathcal{X}$. The pairing $(\cdot\,,\cdot)$ denotes the duality product.

Moreover, it is assumed that the operator $C(x)$ consists of an operator $A(\cdot)$ depending linearly on x and a part B that is independent of x, i.e.,

$$C(x) = A(x) + B \quad \text{and} \quad A(\cdot)u \in \mathcal{L}(\mathcal{X}, \mathcal{Y}_0^*)\,.$$

This assumption already guarantees the unique solvability of (3.73).

The assumption above is for instance satisfied for the parameter estimation problem

$$\begin{aligned} -\operatorname{div}(a\nabla u) + b(u) &= f \quad \text{in } \Omega\,,\\ u &= 0 \quad \text{on } \partial\Omega\,, \end{aligned}$$

with $A(\cdot)$ and B defined by

$$(A(a)u, v) := \int_\Omega a(s)\nabla u \nabla v\, ds\,, \qquad (Bu, v) := \int_\Omega b(u) v\, ds\,,$$

and $\mathcal{X} = \mathcal{Y} = H^1(\Omega)$, $\mathcal{Y}_0 := H_0^1(\Omega)$, $\mathcal{D}(F) = \{a \in \mathcal{X} : a \ge \underline{a} \text{ a.e.}\}$, $\underline{a} > 0$.

In [108], the following modification of Landweber iteration is suggested (compare (3.46)):

$$x_{k+1}^\delta = x_k^\delta + \omega L(x_k^\delta)^*(y^\delta - F(x_k^\delta)) + \beta_k(x_0 - x_k^\delta)\,. \tag{3.74}$$

The linear operator $L(\cdot) : \mathcal{X} \to \mathcal{Y}_0$ is defined by

$$L(x)h = -JA(h)F(x)\,,$$

where $J : \mathcal{Y}_0^* \to \mathcal{Y}_0$ is the duality map. ω is the usual scaling parameter (cf., e.g., (2.24)) and the parameters β_k satisfy the condition

$$0 \le \beta_{k+1} \le \beta_k \le 1 \qquad \text{and} \qquad \beta_k \to 0 \quad \text{as} \quad k \to \infty\,.$$

As for classical Landweber iteration the method (3.74) can only converge if the iteration operator $L(\cdot)^*$ is (locally) uniformly bounded and if the scaling parameter ω is properly chosen.

As stopping rule the discrepancy principle (2.2) is used. However, note that in this modified case in general the data have to satisfy stronger conditions: for a parameter estimation problem as the one above, usually the observation space is chosen to be $L^2(\Omega)$. In this modification, the space $\mathcal{Y} = H^1(\Omega)$, meaning that the data have to be smooth and that the noise can be measured in the norm of the space $H^1(\Omega)$, otherwise the method in (3.74) is not well defined.

Following the ideas in [144] (see Section 3.2), convergence of the iterates $x_{k_*}^\delta$ was proven in [108] provided that condition (3.61) is satisfied, i.e.,

$$\sum_{k=0}^{\infty} \beta_k < \infty\,.$$

To obtain convergence rates appropriate source and smallness conditions have to hold as in the classical case. To obtain the rate $O(\sqrt{\delta})$ in the classical case it is needed that $x^\dagger - x_0 \in \mathcal{R}((F'(x^\dagger)^*F'(x^\dagger))^{\frac{1}{2}}) = \mathcal{R}(F'(x^\dagger)^*)$. It turns out that the appropriate source condition for the modified method (3.74) is given by

$$x^\dagger - x_0 \in \mathcal{R}(L(x^\dagger)^*)\,.$$

If $\|x_0 - x^\dagger\|$ is sufficiently small, if τ in (2.2) is sufficiently large, and if the parameters β_k satisfy a condition similar to (3.68), then it is proven that

$$\|x_k^\delta - x^\dagger\| = O(\sqrt{\beta_{k_*}})\,.$$

For an a-priori stopping rule even the rate $O(\sqrt{\delta})$ could be shown (see [108, Theorem 5.2.1] for details). As already mentioned in Remark 3.18, the condition on the parameters β_k needed for the convergence rates results implies that

$$\sum_{k=0}^{\infty} \beta_k = \infty$$

and hence contradicts condition (3.61) needed for the convergence proof.

3.4 Steepest descent and minimal error method

In this section we consider iterative regularization methods of the form

$$x_{k+1}^\delta = x_k^\delta + \omega_k^\delta s_k^\delta\,, \qquad s_k^\delta := F'(x_k^\delta)^*(y^\delta - F(x_k^\delta))\,, \qquad k \in \mathbb{N}_0\,, \tag{3.75}$$

where the coefficients ω_k^δ are chosen as

$$\omega_k^\delta := \frac{\|s_k^\delta\|^2}{\|F'(x_k^\delta)s_k^\delta\|^2} \tag{3.76}$$

for the *steepest descent method* and

$$\omega_k^\delta := \frac{\|y^\delta - F(x_k^\delta)\|^2}{\|s_k^\delta\|^2} \tag{3.77}$$

for the *minimal error method.* Note that the choice $\omega_k^\delta = 1$ corresponds to the classical Landweber iteration.

In [44] it has been shown that even for the solution of *linear* ill-posed problems the steepest descent method is only a regularization method when stopped via a discrepancy principle and not via an a-priori parameter choice strategy. Therefore, we will use (2.2), (2.14) as stopping rule.

In the following proposition we will prove monotonicity of the errors and well definedness of the steepest descent and minimal error method.

Proposition 3.20 *Assume that* (2.4) *holds and that* $F(x) = y$ *has a solution* $x_* \in \mathcal{B}_\rho(x_0)$. *Let* $k_* = k_*(\delta, y^\delta)$ *be chosen according to the stopping rule* (2.2), (2.14). *Then* x_k^δ *as in* (3.75) *with* ω_k^δ *as in* (3.76) *or* (3.77) *is well defined and*

$$\|x_{k+1}^\delta - x_*\| \le \|x_k^\delta - x_*\|\,, \qquad 0 \le k < k_*\,.$$

Moreover, $x_k^\delta \in \mathcal{B}_\rho(x_*) \subset \mathcal{B}_{2\rho}(x_0)$ *for all* $0 \le k \le k_*$ *and*

$$\sum_{k=0}^{k_*-1} \omega_k^\delta \|y^\delta - F(x_k^\delta)\|^2 \le \Psi^{-1} \|x_0 - x_*\|^2 \tag{3.78}$$

with

$$\Psi := (1 - 2\eta) - 2\tau^{-1}(1 + \eta)\,. \tag{3.79}$$

Proof. Let $x_k^\delta \in \mathcal{B}_\rho(x_*)$ for some $0 \le k < k_*$. To prove that x_{k+1}^δ is well defined, we have to show that $s_k^\delta \neq 0$ and that $F'(x_k^\delta)s_k^\delta \neq 0$.

Let us first assume that $s_k^\delta = 0$. Then (3.75) implies that

$$\begin{aligned} 0 = \langle s_k^\delta, x_k^\delta - x_* \rangle &= \langle y^\delta - F(x_k^\delta), y^\delta - y \rangle - \|y^\delta - F(x_k^\delta)\|^2 \\ &\quad - \langle y^\delta - F(x_k^\delta), F(x_k^\delta) - F(x_*) - F'(x_k^\delta)(x_k^\delta - x_*) \rangle . \end{aligned}$$

Together with (1.2) and (2.4) we now obtain that

$$\|y^\delta - F(x_k^\delta)\|^2 \le \|y^\delta - F(x_k^\delta)\| (\delta + \eta(\delta + \|y^\delta - F(x_k^\delta)\|))$$

and furthermore with (2.14) that

$$\|y^\delta - F(x_k^\delta)\| \le \frac{1+\eta}{1-\eta}\delta \le \frac{\tau}{2}\delta ,$$

which is a contradiction to (2.2). Thus, $s_k^\delta \neq 0$.

Now, assume that $F'(x_k^\delta)s_k^\delta = 0$. Then obviously $s_k^\delta \in \mathcal{N}(F'(x_k^\delta))$. By the definition of s_k^δ (cf. (3.75)) we also have that $s_k^\delta \in \mathcal{R}(F'(x_k^\delta)^*) \subset \mathcal{N}(F'(x_k^\delta))^\perp$. Consequently $s_k^\delta = 0$, which is a contradiction to what we just proved above. This shows that $F'(x_k^\delta)s_k^\delta \neq 0$. Therefore, x_{k+1}^δ with ω_k^δ as in (3.76) or (3.77) is well defined by (3.75) and we obtain with (1.2) and (2.4) that

$$\begin{aligned} &\|x_{k+1}^\delta - x_*\|^2 - \|x_k^\delta - x_*\|^2 \\ &\quad = 2\langle x_{k+1}^\delta - x_k^\delta, x_k^\delta - x_* \rangle + \|x_{k+1}^\delta - x_k^\delta\|^2 \\ &\quad = 2\omega_k^\delta \langle y^\delta - F(x_k^\delta), F'(x_k^\delta)(x_k^\delta - x_*) \rangle + (\omega_k^\delta)^2 \|s_k^\delta\|^2 \\ &\quad \le \omega_k^\delta \|y^\delta - F(x_k^\delta)\| (2(1+\eta)\delta - (1-2\eta) \|y^\delta - F(x_k^\delta)\|) \\ &\qquad - \omega_k^\delta (\|y^\delta - F(x_k^\delta)\|^2 - \omega_k^\delta \|s_k^\delta\|^2 . \end{aligned}$$

Together with (2.2) this yields the estimate

$$\begin{aligned} &\|x_{k+1}^\delta - x_*\|^2 + \Psi \omega_k^\delta \|y^\delta - F(x_k^\delta)\|^2 \\ &\quad \le \|x_k^\delta - x_*\|^2 - \omega_k^\delta (\|y^\delta - F(x_k^\delta)\|^2 - \omega_k^\delta \|s_k^\delta\|^2) \end{aligned} \tag{3.80}$$

with Ψ as in (3.79). Note that, due to (2.14), $\Psi > 0$.

We will now show that the expression in brackets on the right hand side of (3.80) is nonnegative if ω_k^δ is as in (3.76) or (3.77). For the latter it is obvious from the definition that the expression in brackets is equal to zero. Let us now assume that ω_k^δ is defined as in (3.76). Then

$$\omega_k^\delta \|s_k^\delta\|^2 = \frac{\langle F'(x_k^\delta)s_k^\delta, y^\delta - F(x_k^\delta) \rangle^2}{\|F'(x_k^\delta)s_k^\delta\|^2} \le \|y^\delta - F(x_k^\delta)\|^2 .$$

Together with (3.80) this shows that

$$\|x_{k+1}^\delta - x_*\|^2 + \Psi\omega_k^\delta \|y^\delta - F(x_k^\delta)\|^2 \le \|x_k^\delta - x_*\|^2 .$$

Noting that $x_0 \in \mathcal{B}_\rho(x_*)$ the assertions now follow by induction. □

For the classical Landweber iteration we used the scaling properties (2.3) or (2.29) that make the computation of an estimate of $\|F'(x)\|$ necessary. This is no longer needed for the steepest descent or minimal error method. However, to be able to prove convergence we need that F' is continuous and that

$$\omega := \sup_{x \in \mathcal{B}_{2\rho}(x_0)} \|F'(x)\| < \infty . \tag{3.81}$$

This condition guarantees that ω_k^δ defined by either (3.76) or (3.77) is bounded from below by ω^{-2}. Moreover, (2.2) and (3.78) then imply that $k_* < \infty$.

In case of exact data there are two possibilities. If $k_* < \infty$, then an exact solution of $F(x) = y$ has been found. If $k_* = \infty$, then we will show below that the iterates converge to a solution.

Theorem 3.21 *Assume that the conditions* (2.4) *and* (3.81) *hold and that the equation $F(x) = y$ is solvable in $\mathcal{B}_\rho(x_0)$. Let $k_* = k_*(0, y) = \infty$. Then the iterates x_k defined as in* (3.75) *with exact data ($y^\delta = y$) and ω_k as in* (3.76) *or* (3.77) *converge to a solution of $F(x) = y$. If $\mathcal{N}(F'(x^\dagger)) \subset \mathcal{N}(F'(x))$ for all $x \in \mathcal{B}_\rho(x^\dagger)$, then x_k converges to $x^\dagger$ as $k \to \infty$.*

Proof. The proof is analogous to the one of Theorem 2.4. $\|y - F(x_i)\|^2$ has to be replaced by $\omega_i \|y - F(x_i)\|^2$ and we use Proposition 3.20 to conclude that

$$\sum_{k=0}^{\infty} \omega_k \|y - F(x_k)\|^2 < \infty ,$$

which together with (3.81) implies that $\|y - F(x_k)\| \to 0$ as $k \to \infty$. □

The next result shows that the stopping rule (2.2), (2.14) renders the steepest descent and minimal error method a regularization method.

Theorem 3.22 *Assume that the conditions* (2.4) *and* (3.81) *hold and that the equation $F(x) = y$ is solvable in $\mathcal{B}_\rho(x_0)$. Let $k_* = k_*(\delta, y^\delta)$ be chosen according to the stopping rule* (2.2), (2.14)*. Then the iterates $x_{k_*}^\delta$ defined via* (3.75) *with ω_k^δ as in* (3.76) *or* (3.77) *converge to a solution of $F(x) = y$. If $\mathcal{N}(F'(x^\dagger)) \subset \mathcal{N}(F'(x))$ for all $x \in \mathcal{B}_\rho(x^\dagger)$, then $x_{k_*}^\delta$ converges to $x^\dagger$ as $\delta \to 0$.*

Proof. First of all, we want to mention that it is an immediate consequence of (2.2), (3.75) – (3.77), and Proposition 3.20 that for all $0 \le k < k_*(0,y)$ it holds that for $\delta > 0$ sufficiently small

$$\begin{aligned} k_*(\delta, y^\delta) > k\,, \quad x_l^\delta \text{ is well defined}\,, \quad \text{and} \\ \lim_{\delta \to 0} x_l^\delta = x_l\,, \quad 0 \le l \le k+1\,. \end{aligned} \tag{3.82}$$

Let now x_* be the limit of x_k (cf. Theorem 3.21) and let $\{\delta_n\}$ be a sequence converging to zero as $n \to \infty$. Denote by $y_n := y^{\delta_n}$ a corresponding sequence of perturbed data, and by $k_n = k_*(\delta_n, y_n)$ the stopping index determined from the discrepancy principle for the Landweber iteration applied to the pair (δ_n, y_n).

If $\tilde{k} := k_*(0,y) < \infty$, then (3.82) and Proposition 3.20 imply that $k_n \ge \tilde{k}$ for n sufficiently large and that

$$\|x_{k_n}^{\delta_n} - x_*\| \le \|x_{\tilde{k}}^{\delta_n} - x_*\| = \|x_{\tilde{k}}^{\delta_n} - x_{\tilde{k}}\| \to 0 \quad \text{as} \quad n \to \infty\,.$$

If $k_*(0,y) = \infty$, then $k_n \to \infty$ as $n \to \infty$ and the assertion follows as in the proof of Theorem 2.6. □

For linear ill-posed problems, i.e., when $F(x) \equiv F$, the basic idea to prove convergence rates for the steepest descent and minimal error method is to verify that $(F^*F)^{-\frac{1}{2}}(x_k - x^\dagger)$ is bounded if $x^\dagger - x_0 = (F^*F)^{\frac{1}{2}}v$ (cf. [50]). This idea can be carried over to nonlinear problems:

One can show that $(F'(x^\dagger)^*F'(x^\dagger))^{-\frac{1}{2}}(x_k - x^\dagger)$ is bounded if the source condition (2.22) with $\mu = 1/2$ holds and if (2.23) is satisfied.

Theorem 3.23 *Assume that the conditions* (2.23) *with c sufficiently small and* (2.22) *with $\mu = 1/2$ hold and that $x^\dagger \in \mathcal{B}_\rho(x_0)$. Then*

$$\|x_k - x^\dagger\| = O(k^{-\frac{1}{2}})\,,$$

where x_k is defined as in (3.75) *with exact data ($y^\delta = y$) and ω_k as in* (3.76) *or* (3.77).

Proof. See [125]. □

So far, convergence rates have been proven only in the case of exact data.

Remark 3.24 For the solution of linear ill-posed problems it is well known that a reduction of iteration steps may be achieved if instead of (3.75) one uses an iteration procedure, where the search direction is different from the gradient direction, i.e.,

$$x_{k+1}^\delta = x_k^\delta + \omega_k^\delta s_k^\delta\,, \qquad s_k^\delta := F'(x_k^\delta)^*(y^\delta - F(x_k^\delta)) + \beta_{k-1}^\delta s_{k-1}^\delta\,, \tag{3.83}$$

$k \in \mathbb{N}_0$. The coefficients ω_k^δ and β_{k-1}^δ depend on $k, x_k^\delta, x_{k-1}^\delta, \dots, x_0, y^\delta$. The most prominent representative is the *conjugate gradient method*.

When using iteration methods (3.83) for nonlinear ill-posed problems, one can prove similar results as for the steepest descent and minimal error method provided that the coefficients ω_k^δ and β_{k-1}^δ depend continuoulsy on the iterates and satisfy some additional conditions (cf. [142]). Unfortunately, conjugate gradient type methods do not satisfy these conditions and to our knowledge no convergence results exist for these methods when applied to nonlinear ill-posed problems.

3.5 The Landweber–Kaczmarz method

In this section we study the numerical solution of an operator equation $F(x) = y$ where

$$F := (F_0, \dots, F_{p-1}) : \bigcap_{i=0}^{p-1} \mathcal{D}(F_i) \subset \mathcal{X} \to \mathcal{Y}^p \tag{3.84}$$

and $y = (y_0, \dots, y_{p-1})$ with $p > 1$. For this problem the Landweber iteration reads as follows

$$x_{k+1}^\delta = x_k^\delta + \sum_{i=0}^{p-1} F_i'(x_k^\delta)^* (y_i^\delta - F_i(x_k^\delta)) , \qquad k \in \mathbb{N}_0 .$$

The *Landweber–Kaczmarz method* is defined by

$$x_{k+1}^\delta = x_k^\delta + F_k'(x_k^\delta)^* (y_k^\delta - F_k(x_k^\delta)) , \qquad k \in \mathbb{N}_0 . \tag{3.85}$$

For simplicity of presentation, here and below we use the notation

$$F_k := F_{r_k} \quad \text{and} \quad y_k^\delta := y_{r_k}^\delta \quad \text{with} \quad k = p \lfloor k/p \rfloor + r_k \quad \text{and} \quad r_k \in \{0, \dots, p-1\} . \tag{3.86}$$

With $\lfloor x \rfloor$ we denote the biggest integer less or equal x, i.e., $\lfloor 1 \rfloor = 1$, $\lfloor 0.9 \rfloor = 0$.

The Landweber–Kaczmarz method applies the Landweber iteration steps cyclically. It is obvious that the principal idea of Kaczmarz's method, also called algebraic reconstruction technique (ART) (cf. [69]), can be used in conjunction with any other iterative procedure, such as Newton type methods (cf. [22]).

The Kaczmarz algorithm has proven to be an efficient method for solving inverse problems with a bilinear structure (see, e.g., [122]). It is closely related to the method of *adjoint fields* frequently used by engineers (see [19, 149]). Some more references on Kaczmarz type algorithms are [6, 112, 113, 115].

In order to prove convergence and stability of the Landweber–Kaczmarz method (cf. [62, 101]), we basically require the same conditions on the operators F_i as they are required for F for proving convergence and stability of Landweber iteration, i.e., we assume that

$$\mathcal{B}_{2\rho}(x_0) \subset \bigcap_{i=0}^{p-1} \mathcal{D}(F_i) \quad \text{for some} \quad \rho > 0\,, \tag{3.87}$$

that the operators F_i are Fréchet-differentiable in the closed ball $\mathcal{B}_{2\rho}(x_0)$, and that each of them satisfies the conditions (2.3) and (2.4).

In particular, it follows from these assumptions that (2.5) holds for every F_i, $i = 0, \ldots, p-1$. This, together with $\mathcal{N}(F'(x)) = \bigcap_{i=0}^{p-1} \mathcal{N}(F_i'(x))$ implies that

$$F(x) = F(\tilde{x}) \iff x - \tilde{x} \in \mathcal{N}(F'(x))$$

holds for $x, \tilde{x} \in \mathcal{B}_{2\rho}(x_0)$. Under the above assumptions it follows that

$$\tfrac{1}{p} \sum_{i=0}^{p-1} \|F_i'(x)\|^2 \le 1$$

for all $x \in \mathcal{B}_{2\rho}(x_0)$ and that

$$\begin{aligned} \|F(x) - F(\tilde{x}) - F'(x)(x-\tilde{x})\|^2 & \\ &= \sum_{i=0}^{p-1} \|F_i(x) - F_i(\tilde{x}) - F_i'(x)(x-\tilde{x})\|^2 \\ &\le \eta^2 \sum_{i=0}^{p-1} \|F_i(x) - F_i(\tilde{x})\|^2 = \eta^2 \|F(x) - F(\tilde{x})\|^2 \end{aligned}$$

for all $x, \tilde{x} \in \mathcal{B}_{2\rho}(x_0)$. Thus, the operator $F/\sqrt{p}$ satisfies the convergence conditions of Landweber iteration. However, the opposite is not true, i.e., if $F/\sqrt{p}$ satisfies (2.4), then it is not guaranteed that each F_i satisfies (2.4). In this sense the convergence conditions of the Landweber–Kaczmarz method are stronger than for the Landweber iteration.

As for the Landweber iteration we need a stopping rule. Here it is convenient to modify (3.85) as follows

$$x_{k+1}^\delta = x_k^\delta + \omega_k^\delta F_k'(x_k^\delta)^*(y_k^\delta - F_k(x_k^\delta))\,, \qquad k \in \mathbb{N}_0\,, \tag{3.88}$$

where

$$\omega_k^\delta := \begin{cases} 1\,, & \text{if } \tau\delta < \|y_k^\delta - F_k(x_k^\delta)\|\,, \\ 0\,, & \text{else}\,, \end{cases}$$

with τ as in (2.14). The iteration is terminated if it stagnates over a full cycle of p successive iterates, i.e., the iteration is stopped after $k_* = k_*(\delta, y^\delta)$ steps with

$$\omega^\delta_{k_*-i} = 0\,, \quad i = 0, \dots, p-1\,, \qquad \text{and} \qquad \omega^\delta_{k_*-p} = 1\,. \tag{3.89}$$

Note that in case of exact data, the Landweber–Kaczmarz iterates (3.85) and its modifications (3.88) coincide. The stopping rule (3.89) for the modified method guarantees that x_{k_*} is a solution of $F(x) = y$ if $k_* = k_*(0, y)$ is finite.

For comparison the Landweber iteration is stopped according to the discrepancy principle (2.2), which for F as in (3.84) reads as

$$\sum_{i=0}^{p-1} \|y_i^\delta - F_i(x^\delta_{k_*})\|^2 \le p\,\tau^2\delta^2 < \sum_{i=0}^{p-1} \|y_i^\delta - F_i(x^\delta_k)\|^2\,, \qquad 0 \le k < k_*\,.$$

Obviously, for the case $p = 1$, the termination procedure for the modified Landweber–Kaczmarz method above results in Landweber iteration stopped by (2.2). If $p > 1$, then the stopping rule (3.89) guarantees that

$$\|y_i^\delta - F_i(x^\delta_{k_*})\| \le \tau\delta\,, \qquad i = 0, \dots, p-1\,,$$

since then $x^\delta_{k_*} = x^\delta_{k_*-i}$ for all $i = 0, \dots, p-1$.

To prove local convergence, we will always assume that the equation $F(x) = y$ is solvable in $\mathcal{B}_\rho(x_0)$. As already mentioned at the end of Section 2.1, condition (2.5) implies that then there also exists a unique x_0-minimum-norm solution $x^\dagger$ satisfying condition (2.9), i.e.,

$$x^\dagger - x_0 \in \mathcal{N}(F'(x^\dagger))^\perp.$$

Analogously to Proposition 2.2 it can be shown that, if $F(x) = y$ has a solution x_* in $\mathcal{B}_\rho(x_0)$ and if the operators F_i are Fréchet-differentiable in $\mathcal{B}_{2\rho}(x_0)$ and satisfy the conditions (2.3), (2.4), and (3.87), then the errors of the iterates of the Landweber–Kaczmarz method are monotonically decreasing for $0 \le k < k_*$, i.e., under these assumptions it can be shown that, if $x^\delta_k \in \mathcal{B}_\rho(x_*)$, a sufficient condition for x^δ_{k+1} as in (3.85) to be a better approximation of x_* than x^δ_k is that

$$\|y_k^\delta - F_k(x^\delta_k)\| > 2\frac{1+\eta}{1-2\eta}\delta\,.$$

Moreover, it then holds that $x^\delta_k, x^\delta_{k+1} \in \mathcal{B}_\rho(x_*) \subset \mathcal{B}_{2\rho}(x_0)$. Following (2.13), one can analogously show that for $\delta = 0$

$$\|x_{k+1} - x_*\|^2 + (1 - 2\eta)\,\|y_k - F_k(x_k)\|^2 \le \|x_k - x^*\|^2\,.$$

From this it follows by induction that

$$\sum_{k=0}^{\infty} \|y_k - F_k(x_k)\|^2 \leq \frac{1}{(1-2\eta)} \|x_0 - x^*\|^2 . \tag{3.90}$$

We will now prove local convergence of the Landweber–Kaczmarz method (3.85) in case of exact data.

Theorem 3.25 *Assume that the operators F_i are Fréchet-differentiable in $\mathcal{B}_{2\rho}(x_0)$ and satisfy the conditions* (2.3), (2.4), *and* (3.87). *Moreover, we assume that $F(x) = y$ has a a solution in $\mathcal{B}_\rho(x_0)$, where F is as in* (3.84). *Then the Landweber–Kaczmarz method applied to exact data y converges to a solution of $F(x) = y$. If*

$$\mathcal{N}(F_i'(x^\dagger)) \subset \mathcal{N}(F_i'(x)) \quad \text{for all} \quad x \in \mathcal{B}_\rho(x^\dagger), \quad i = 0, 1, \ldots, p-1, \tag{3.91}$$

then x_k converges to $x^\dagger$.

Proof. The proof is similar to the one of Theorem 2.4, but differs at a central point. As mentioned above, we know that the unique x_0-minimum-norm solution $x^\dagger$ exists in $\mathcal{B}_\rho(x_0)$ and that $\{\|e_k\|\}$ is monotonically decreasing, where $e_k := x_k - x^\dagger$. Thus, it converges to some $\varepsilon \geq 0$. For given $k \leq j$ we have

$$k = p\lfloor k/p \rfloor + r_k \leq j = p\lfloor j/p \rfloor + r_j \quad \text{with} \quad r_k, r_j \in \{0, \ldots, p-1\} .$$

Now we can choose $l_0 \in \mathbb{N}_0$ in such a way that $\lfloor k/p \rfloor \leq l_0 \leq \lfloor j/p \rfloor$ and

$$\sum_{s=0}^{p-1} \|y_s - F_s(x_{p\,l_0+s})\| \leq \sum_{s=0}^{p-1} \|y_s - F_s(x_{p\,i_0+s})\| \tag{3.92}$$

for $\lfloor k/p \rfloor \leq i_0 \leq \lfloor j/p \rfloor$. Moreover, we select

$$l := p\,l_0 + (p-1) . \tag{3.93}$$

Again, we have (2.18) and (2.19) and as in the proof of Theorem 2.4 it suffices to show that $\langle e_l - e_j, e_l \rangle$ and $\langle e_l - e_k, e_l \rangle$ tend to zero for $k \to \infty$. Following the lines of the proof of Theorem 2.4 we obtain together with (2.4) and (3.85) that

$$\begin{aligned} |\langle e_l - e_k, e_l \rangle| \;\; &\leq \;\; (1+\eta) \sum_{i=k}^{l-1} \|y_i - F_i(x_i)\| \, \|y_i - F_i(x_l)\| \\ &\quad + 2\eta \sum_{i=k}^{l-1} \|y_i - F_i(x_i)\|^2 . \end{aligned}$$

In particular, from this and (3.90) it follows that $\langle e_l - e_k, e_l \rangle \to 0$ for $k \to \infty$ if

$$\sum_{i=k}^{l-1} \|y_i - F_i(x_i)\| \, \|y_i - F_i(x_l)\| \to 0 \, .$$

To prove this, we will first derive an estimate for $\|y_i - F_i(x_l)\|$: since $i = p \lfloor i/p \rfloor + r_i$ with $r_i \in \{0, \dots, p-1\}$, we obtain with (2.5), (2.3), (3.85), (3.86), and (3.93) that

$$\begin{aligned}
\|y_i - F_i(x_l)\| &= \|y_{r_i} - F_{r_i}(x_{p\, l_0 + p - 1})\| \\
&\le \|y_{r_i} - F_{r_i}(x_{p\, l_0 + r_i})\| + \sum_{s=r_i}^{p-2} \|F_{r_i}(x_{p\, l_0 + s + 1}) - F_{r_i}(x_{p\, l_0 + s})\| \\
&\le \|y_{r_i} - F_{r_i}(x_{p\, l_0 + r_i})\| + \frac{1}{1-\eta} \sum_{s=r_i}^{p-2} \|F'_{r_i}(x_{p\, l_0 + s})(x_{p\, l_0 + s + 1} - x_{p\, l_0 + s})\| \\
&\le \|y_{r_i} - F_{r_i}(x_{p\, l_0 + r_i})\| + \frac{1}{1-\eta} \sum_{s=r_i}^{p-2} \|x_{p\, l_0 + s + 1} - x_{p\, l_0 + s}\| \\
&= \|y_{r_i} - F_{r_i}(x_{p\, l_0 + r_i})\| + \frac{1}{1-\eta} \sum_{s=r_i}^{p-2} \|F'_s(x_{p\, l_0 + s})^* (y_s - F_s(x_{p\, l_0 + s}))\| \\
&\le \frac{2-\eta}{1-\eta} \sum_{s=r_i}^{p-2} \|y_s - F_s(x_{p\, l_0 + s})\| \, .
\end{aligned}$$

Together with (3.92) this yields the estimate

$$\begin{aligned}
&\sum_{i=k}^{l-1} \|y_i - F_i(x_i)\| \, \|y_i - F_i(x_l)\| \\
&\quad \le \sum_{i_0 = \lfloor k/p \rfloor}^{l_0} \sum_{r_i = 0}^{p-1} \|y_{r_i} - F_{r_i}(x_{p\, i_0 + r_i})\| \, \|y_{r_i} - F_{r_i}(x_l)\| \\
&\quad \le \frac{2-\eta}{1-\eta} \sum_{i_0 = \lfloor k/p \rfloor}^{l_0} \Big(\sum_{r_i = 0}^{p-1} \|y_{r_i} - F_{r_i}(x_{p\, i_0 + r_i})\| \sum_{s=r_i}^{p-2} \|y_s - F_s(x_{p\, l_0 + s})\| \Big) \\
&\quad \le \frac{2-\eta}{1-\eta} \sum_{i_0 = \lfloor k/p \rfloor}^{l_0} \Big(\sum_{r_i = 0}^{p-1} \|y_{r_i} - F_{r_i}(x_{p\, i_0 + r_i})\| \Big)^2
\end{aligned}$$

$$\leq \frac{2-\eta}{1-\eta} p \sum_{i_0=\lfloor k/p \rfloor}^{l_0} \sum_{r_i=0}^{p-1} \|y_{r_i} - F_{r_i}(x_{p\,i_0+r_i})\|^2$$

$$\leq \frac{2-\eta}{1-\eta} p \sum_{i=k-r_k}^{l} \|y_i - F_i(x_i)\|^2 .$$

Since $k - r_k \to \infty$ if $k \to \infty$, it now follows together with (3.90) that $\langle e_l - e_k, e_l \rangle \to 0$ for $k \to \infty$. Analogously one can show that $\langle e_l - e_j, e_l \rangle \to 0$.

Obviously (3.90) implies that $\|y_k - F_k(x_k)\| \to 0$ as $k \to \infty$. Hence, (3.86) yields that

$$\lim_{n\to\infty} \|y_s - F_s(x_{p\,n+s})\| = 0 \qquad \text{for all} \qquad s \in \{0,\dots,p-1\} .$$

Now a similar argument as in the proof of Theorem 2.4 shows that $\{x_k\}$ converges to a solution of $F(x) = y$.

If (3.91) holds, then by definition of the Landweber–Kaczmarz method (cf. (3.85)) it follows with (3.86) that

$$x_{k+1} - x_k \in \mathcal{R}(F_k'(x_k)^*) \subset \mathcal{N}(F_k'(x_k))^\perp \subset \mathcal{N}(F_k'(x^\dagger))^\perp \subset \Big(\bigcap_{i=0}^{p-1} \mathcal{N}(F_i'(x^\dagger))\Big)^\perp$$

and hence that

$$x_k - x_0 \in \Big(\bigcap_{i=0}^{p-1} \mathcal{N}(F_i'(x^\dagger))\Big)^\perp \quad \text{for all} \quad k \in \mathbb{N} .$$

Therefore, this then also holds for the limit of x_k. Since $x^\dagger$ is the unique solution for which this condition holds (cf. (2.9)), this proves that $x_k \to x^\dagger$ as $k \to \infty$. □

The next theorem shows that the modified Landweber–Kaczmarz method (3.88) in combination with the discrepancy principle is a regularization method.

Theorem 3.26 *Let the assumptions of Theorem* 3.25 *hold and assume that* $\|y_i^\delta - y_i\| \leq \delta$ *for all* $i \in \{0,\dots,p-1\}$. *Moreover, let* $k_* = k_*(\delta, y^\delta)$ *be determined according to* (3.89). *Then the modified Landweber–Kaczmarz iterates* $x_{k_*}^\delta$ *converge to a solution of* $F(x) = y$. *If in addition* (3.91) *holds, then* $x_{k_*}^\delta$ *converges to* $x^\dagger$ *as* $\delta \to 0$.

Proof. The proof is analogous to the one of Theorem 2.6. □

4 Newton type methods

As can be seen from the previous chapters, Landweber iteration and its modifications are easy to realize numerically. However, the number of iterations needed can be rather large. Therefore, one wants to find faster methods. For well-posed problems Newton type methods are the appropriate answer. The key idea of any Newton type method consists in repeatedly linearizing the operator equation $F(x) = y$ around some approximate solution x_k^δ, and then solving the linearized problem

$$F'(x_k^\delta)(x_{k+1}^\delta - x_k^\delta) = y^\delta - F(x_k^\delta)$$

for x_{k+1}^δ. However, usually these linearized problems are also ill-posed if the nonlinear problem is ill-posed (cf. [45, Section 10.1]) and, therefore, they have to be regularized.

If we apply Tikhonov regularization to the linearized problem, we end up with the Levenberg–Marquardt method that is treated in the first section of this chapter. Adding a penalty term yields the well-known iteratively regularized Gauss–Newton method introduced by Bakushinskii (see Section 4.2). Generalizations of this method, where instead of Tikhonov regularization other regularization techniques are used, can be found in Section 4.3.

One disadvantage of these Newton type methods is that usually operators involving the derivative of F have to be inverted. For well-posed problems a reduction of the numerical effort is achieved via Quasi-Newton methods, e.g., Broyden's method. In Section 4.4, a modification of this method is investigated that may be applied to ill-posed problems.

4.1 Levenberg–Marquardt method

The original idea of the *Levenberg–Marquardt method* was to minimize the squared misfit, $\|F(x) - y^\delta\|^2$, within a trust region, i.e., a ball of radius η_k around x_k^δ is chosen and the linearized functional is minimized within this ball. This is easily seen to be equivalent to minimizing

$$\|y^\delta - F(x_k^\delta) - F'(x_k^\delta)z\|^2 + \alpha_k \|z\|^2 \tag{4.1}$$

for $z = z_k$, where α_k is the corresponding Lagrange parameter. Then this is repeated with $x_{k+1}^\delta = x_k^\delta + z_k$ instead of x_k^δ and possibly some updated trust region radius η_{k+1} until convergence.

Another justification for (4.1) is the regularization induced by adding the penalty term $\alpha_k \|z\|^2$ to the linearized problem. This is equivalent to applying Tikhonov regularization to the linearized problem $F'(x_k^\delta)h = y^\delta - F(x_k^\delta)$ yielding the iteration procedure

$$x_{k+1}^\delta = x_k^\delta + (F'(x_k^\delta)^* F'(x_k^\delta) + \alpha_k I)^{-1} F'(x_k^\delta)^* (y^\delta - F(x_k^\delta)), \tag{4.2}$$

where as above y^δ are noisy data satisfying the estimate (1.2). The sequence of iteration dependent regularization parameters α_k has to be chosen appropriately. Hanke [63] proposes to do so a-posteriori by the following discrepancy principle: let $\alpha = \alpha_k$ be such that

$$p_k^\delta(\alpha) := \|y^\delta - F(x_k^\delta) - F'(x_k^\delta)(x_{k+1}^\delta(\alpha) - x_k^\delta)\| = q\,\|y^\delta - F(x_k^\delta)\| \tag{4.3}$$

with some fixed $q \in (0,1)$, where $x_{k+1}^\delta(\alpha)$ is defined as in (4.2) with α_k replaced by α. Since

$$p_k^\delta(\alpha) = \alpha\,\|(F'(x_k^\delta)F'(x_k^\delta)^* + \alpha I)^{-1}(y^\delta - F(x_k^\delta))\|,$$

it immediately follows that the function p_k^δ is continuous and strictly increasing if $y^\delta - F(x_k^\delta) \neq 0$ and if F' is continuous, which we assume througout this section. Moreover,

$$\begin{aligned} \lim_{\alpha\to\infty} p_k^\delta(\alpha) &= \|y^\delta - F(x_k^\delta)\|, \\ \lim_{\alpha\to 0} p_k^\delta(\alpha) &= \|P_k^\delta(y^\delta - F(x_k^\delta))\| \le \|y^\delta - F(x_k^\delta) - F'(x_k^\delta)(x - x_k^\delta)\| \end{aligned}$$

for any $x \in \mathcal{D}(F)$, where P_k^δ is the orthogonal projector onto $\mathcal{R}(F'(x_k^\delta))^\perp$. Therefore, (4.3) has a unique solution α_k provided

$$\|y^\delta - F(x_k^\delta) - F'(x_k^\delta)(x^\dagger - x_k^\delta)\| \le \tfrac{q}{\gamma}\,\|y^\delta - F(x_k^\delta)\| \tag{4.4}$$

holds for some $\gamma > 1$, where $x^\dagger$ denotes the x_0-minimum-norm solution (see the end of Section 2.1), which is assumed to exist in $\mathcal{B}_\rho(x_0)$ in the following. It even holds that, among all $x \in \mathcal{X}$ with $\|y^\delta - F(x_k^\delta) - F'(x_k^\delta)(x - x_k^\delta)\| \le q\,\|y^\delta - F(x_k^\delta)\|$, $x = x_{k+1}^\delta$ with α_k determined by (4.3) is the one with minimal distance to x_k^δ (cf. Groetsch [53, p. 44]).

Following Hanke [63], we will now state some convergence results for this method. First of all, similarly to Proposition 2.2 for nonlinear Landweber iteration, monotonicity of the error can be established.

Proposition 4.1 *Let* $0 < q < 1 < \gamma$ *and assume that* (1.1) *has a solution and that* (4.4) *holds so that* α_k *can be defined via* (4.3). *Then, the following estimates hold:*

$$\|x_k^\delta - x^\dagger\|^2 - \|x_{k+1}^\delta - x^\dagger\|^2 \geq \|x_{k+1}^\delta - x_k^\delta\|^2, \tag{4.5}$$

$$\begin{aligned} &\|x_k^\delta - x^\dagger\|^2 - \|x_{k+1}^\delta - x^\dagger\|^2 \\ &\quad\geq \frac{2(\gamma-1)}{\gamma\alpha_k}\|y^\delta - F(x_k^\delta) - F'(x_k^\delta)(x_{k+1}^\delta - x_k^\delta)\|^2 \qquad (4.6)\\ &\quad\geq \frac{2(\gamma-1)(1-q)q}{\gamma\|F'(x_k^\delta)\|^2}\|y^\delta - F(x_k^\delta)\|^2. \qquad (4.7) \end{aligned}$$

Proof. Setting $K_k := F'(x_k^\delta)$ and using (4.2), we obtain analogously to estimate (2.13) for Landweber iteration that

$$\begin{aligned} &\|x_{k+1}^\delta - x^\dagger\|^2 - \|x_k^\delta - x^\dagger\|^2 \\ &\quad= 2\langle x_{k+1}^\delta - x_k^\delta, x_k^\delta - x^\dagger \rangle + \|x_{k+1}^\delta - x_k^\delta\|^2 \\ &\quad= \langle (K_k K_k^* + \alpha_k I)^{-1}(y^\delta - F(x_k^\delta)), 2K_k(x_k^\delta - x^\dagger) \\ &\qquad + (K_k K_k^* + \alpha_k I)^{-1} K_k K_k^*(y^\delta - F(x_k^\delta)) \rangle \\ &\quad= -2\alpha_k\|(K_k K_k^* + \alpha_k I)^{-1}(y^\delta - F(x_k^\delta))\|^2 \\ &\qquad - \|(K_k^* K_k + \alpha_k I)^{-1} K_k^*(y^\delta - F(x_k^\delta))\|^2 \\ &\qquad + 2\langle (K_k K_k^* + \alpha_k I)^{-1}(y^\delta - F(x_k^\delta)), y^\delta - F(x_k^\delta) - K_k(x^\dagger - x_k^\delta) \rangle \\ &\quad\leq -\|x_{k+1}^\delta - x_k^\delta\|^2 - 2\alpha_k^{-1}\|y^\delta - F(x_k^\delta) - K_k(x_{k+1}^\delta - x_k^\delta)\| \qquad (4.8)\\ &\qquad \Big(\|y^\delta - F(x_k^\delta) - K_k(x_{k+1}^\delta - x_k^\delta)\| - \|y^\delta - F(x_k^\delta) - K_k(x^\dagger - x_k^\delta)\|\Big), \end{aligned}$$

where we have used the Cauchy–Schwarz inequality and the identity

$$\alpha_k(K_k K_k^* + \alpha_k I)^{-1}(y^\delta - F(x_k^\delta)) = y^\delta - F(x_k^\delta) - K_k(x_{k+1}^\delta - x_k^\delta). \tag{4.9}$$

By (4.4) and the parameter choice (4.3), we have

$$\|y^\delta - F(x_k^\delta) - K_k(x^\dagger - x_k^\delta)\| \leq \gamma^{-1}\|y^\delta - F(x_k^\delta) - K_k(x_{k+1}^\delta - x_k^\delta)\|.$$

Thus, (4.8) and $\gamma > 1$ imply (4.5) and (4.6).

To obtain, (4.7), we derive an upper bound for α_k: it follows immediately from (4.3) and (4.9) that

$$\begin{aligned} q\|y^\delta - F(x_k^\delta)\| &= \alpha_k\|(K_k K_k^* + \alpha_k I)^{-1}(y^\delta - F(x_k^\delta))\| \\ &\geq \frac{\alpha_k}{\alpha_k + \|K_k\|^2}\|y^\delta - F(x_k^\delta)\| \end{aligned}$$

and hence that

$$\alpha_k \le \frac{q}{1-q} \|K_k\|^2 .$$

This together with (4.3) yields (4.7). □

Based on this proposition, and similarly to Theorem 2.4, convergence in case of exact data can be proven if condition (2.7) holds in a closed ball around x_0, i.e.,

$$\begin{aligned} &\|F(x) - F(\tilde{x}) - F'(x)(x - \tilde{x})\| \le c\|x - \tilde{x}\| \, \|F(x) - F(\tilde{x})\| \, , \\ &x, \tilde{x} \in \mathcal{B}_{2\rho}(x_0) \subset \mathcal{D}(F) \, . \end{aligned} \tag{4.10}$$

Note that according to Proposition 2.1 this already yields the unique existence of the x_0-minimum-norm solution $x^\dagger \in \mathcal{B}_\rho(x_0)$ if the equation (1.1) is solvable in $\mathcal{B}_\rho(x_0)$.

Theorem 4.2 *Let* $0 < q < 1$ *and assume that* (1.1) *is solvable in* $\mathcal{B}_\rho(x_0)$*, that* F' *is uniformly bounded in* $\mathcal{B}_\rho(x^\dagger)$*, and that the Taylor remainder of* F *satisfies* (4.10) *for some* $c > 0$*. Then the Levenberg–Marquardt method with exact data* $y^\delta = y$*,* $\|x_0 - x^\dagger\| < q/c$ *and* α_k *determined from* (4.3)*, converges to a solution of* $F(x) = y$ *as* $k \to \infty$*.*

Proof. Define $\gamma := q(c\|x_0 - x^\dagger\|)^{-1}$ which is greater than 1 by assumption. Then (4.10) with $x = x_0$ and $\tilde{x} = x^\dagger$ implies (4.4), and hence, by Proposition 4.1,

$$\|x_{k+1} - x^\dagger\| \le \|x_k - x^\dagger\|$$

for $k = 0$. By induction this inequality remains true for all $k \in \mathbb{N}$ showing that during the entire iteration the assumptions of Proposition 4.1 are satisfied and $\|x_k - x^\dagger\|$ is monotonically decreasing .

The rest of the proof is similarly to the proof of Theorem 2.4 for Landweber iteration: for arbitrary $j \ge k$ we choose l between k and j according to (2.17). Hence, setting $K_k := F'(x_k)$ and $\eta := c\|x_0 - x^\dagger\| \ge c\|x_i - x^\dagger\|$ for $i \ge 0$, we obtain together with (4.10) that

$$\begin{aligned} \|K_i(x_l - x^\dagger)\| &\le \|y - F(x_l)\| + c\|x_i - x^\dagger\| \, \|y - F(x_i)\| \\ &\quad + c\|x_l - x_i\| \, \|F(x_l) - F(x_i)\| \\ &\le (1 + 5\eta) \|y - F(x_i)\| \, . \end{aligned}$$

Setting $e_k := x_k - x^\dagger$, we now obtain together with (4.9) that

$$\begin{aligned}
|\langle e_l - e_j, e_l \rangle| &= \Big| \sum_{i=l}^{j-1} \langle (K_i^* K_i + \alpha_i I)^{-1} K_i^* (y - F(x_i)), e_l \rangle \Big| \\
&\le \sum_{i=l}^{j-1} \|(K_i K_i^* + \alpha_i I)^{-1}(y - F(x_i))\| \, \|K_i e_l\| \\
&\le (1 + 5\eta) \sum_{i=l}^{j-1} \alpha_i^{-1} \|y - F(x_i) - K_i(x_{i+1} - x_i)\| \, \|y - F(x_i)\| .
\end{aligned}$$

Due to (4.3) and (4.6), the right hand side is bounded from above by

$$\frac{\gamma(1+5\eta)}{2q(\gamma-1)} (\|e_l\|^2 - \|e_j\|^2) .$$

Similarly, one can show that

$$|\langle e_l - e_k, e_l \rangle| \le \frac{\gamma(1+5\eta)}{2q(\gamma-1)} (\|e_k\|^2 - \|e_l\|^2) .$$

Consequently, by (2.18), (2.19), and the fact that the monotone sequence $\{\|e_k\|\}$ has a limit, we can conclude that $\{e_k\}$ and hence $\{x_k\}$ is a Cauchy sequence. Since F' is uniformly bounded in $\mathcal{B}_\rho(x^\dagger)$, by (4.7), $\|y - F(x_k)\|$ goes to zero as $k \to \infty$, so the limit of x_k has to be a solution of (1.1). □

For noisy data the iteration has to be stopped after an appropriate number of steps. If this index k_* is determined by the discrepancy principle (2.2), i.e.,

$$\|y^\delta - F(x_{k_*}^\delta)\| \le \tau\delta < \|y^\delta - F(x_k^\delta)\| , \qquad 0 \le k < k_* ,$$

with the constant τ chosen larger than $1/q$, then we obtain convergence of the iterates $x_{k_*}^\delta$ towards a solution of (1.1) as the noise level δ tends to zero.

Theorem 4.3 *Let the assumptions of Theorem* 4.2 *hold. Additionally let* $k_* = k_*(\delta, y^\delta)$ *be chosen according to the stopping rule* (2.2) *with* $\tau > 1/q$. *Then for* $\|x_0 - x^\dagger\|$ *sufficiently small, the discrepancy principle* (2.2) *terminates the Levenberg–Marquardt method with* α_k *determined from* (4.3) *after finitely many iterations* k_**, and*

$$k_*(\delta, y^\delta) = O(1 + |\ln \delta|) .$$

Moreover, the Levenberg–Marquardt iterates $x_{k_*}^\delta$ *converge to a solution of the equation* $F(x) = y$ *as* $\delta \to 0$.

Proof. First of all, we show monotonicity of $\|x_k^\delta - x^\dagger\|$ up to the stopping index defined by (2.2). If $k_* = 0$, nothing has to be shown. Let us now assume that $k_* \geq 1$ and that

$$\|x_0 - x^\dagger\| < \frac{q\tau - 1}{c(1+\tau)}$$

with c as in (4.10), which by

$$\begin{aligned}\|y^\delta - F(x_0) - F'(x_0)(x^\dagger - x_0)\| &\leq \delta + c\|x_0 - x^\dagger\|\,\|y - F(x_0)\|\\ &\leq (1 + c\|x_0 - x^\dagger\|)\delta + c\|x_0 - x^\dagger\|\,\|y^\delta - F(x_0)\|\end{aligned}$$

implies that (4.4) holds with $\gamma = q\tau[1 + c(1+\tau)\|x_0 - x^\dagger\|]^{-1} > 1$. Hence, we can apply Proposition 4.1 for $k = 0$. An induction argument as in the proof of Theorem 4.2 now yields the desired monotonicity result.

Summing up both sides of (4.7) from 0 to $k_* - 1$, one obtains with (2.2) that

$$\begin{aligned}k_*\tau^2\delta^2 &\leq \sum_{k=0}^{k_*-1} \|y^\delta - F(x_k^\delta)\|^2\\ &\leq \frac{\gamma}{2(\gamma-1)(1-q)q} \sup_{x\in\mathcal{B}_\rho(x^\dagger)} \|F'(x)\|^2\,\|x_0 - x^\dagger\|^2.\end{aligned}$$

Thus, k_* is finite for $\delta > 0$.

Convergence of $x_{k_*}^\delta$ to a solution of (1.1) follows now analogously to the proof of Theorem 2.6 for Landweber iteration.

To show the logarithmic estimate for k_*, we use the triangle inequality, assumption (4.10) and (4.3) to conclude that

$$\begin{aligned}q\|y^\delta - F(x_k^\delta)\| &= \|y^\delta - F(x_k^\delta) - F'(x_k^\delta)(x_{k+1}^\delta - x_k^\delta)\|\\ &\geq \|y^\delta - F(x_{k+1}^\delta)\| - \|F(x_{k+1}^\delta) - F(x_k^\delta) - F'(x_k^\delta)(x_{k+1}^\delta - x_k^\delta)\|\\ &\geq \|y^\delta - F(x_{k+1}^\delta)\| - c\|x_{k+1}^\delta - x_k^\delta\|\,\|F(x_{k+1}^\delta) - F(x_k^\delta)\|\\ &\geq (1 - c\|x_{k+1}^\delta - x_k^\delta\|)\|y^\delta - F(x_{k+1}^\delta)\| - c\|x_{k+1}^\delta - x_k^\delta\|\,\|y^\delta - F(x_k^\delta)\|\end{aligned}$$

and hence with (2.2) and (4.5) that

$$\tau\delta \leq \|y^\delta - F(x_{k_*-1}^\delta)\| \leq \tilde{q}\|y^\delta - F(x_{k_*-2}^\delta)\| \leq \tilde{q}^{k_*-1}\|y^\delta - F(x_0)\|$$

with

$$\tilde{q} := \frac{q + c\|x_0 - x^\dagger\|}{1 - c\|x_0 - x^\dagger\|}.$$

If $\|x_0 - x^\dagger\|$ is sufficiently small, then $\tilde{q} < 1$, which yields the desired estimate for k_*. □

Convergence rates under source conditions (2.22), i.e.,

$$x^\dagger - x_0 = (F'(x^\dagger)^* F'(x^\dagger))^\mu v\,, \qquad \mu > 0\,, \quad v \in \mathcal{N}(F'(x^\dagger))^\perp,$$

have been established by Rieder in [137] (see also [138]) for

$$0 < \mu_{\min} \le \mu \le \tfrac{1}{2}\,, \tag{4.11}$$

under the condition

$$\begin{aligned} F'(x) = R_x F'(x^\dagger) \quad \text{and} \quad \|I - R_x\| \le c_R \|x - x^\dagger\|\,,& \\ x \in \mathcal{B}_\rho(x_0) \subset \mathcal{D}(F)\,,& \end{aligned} \tag{4.12}$$

which is related to Assumption 2.7 (ii) in Section 2.3 above. The minimal exponent $\mu_{\min}$ depends on the constant c_R in (4.12) and is strictly positive. Saturation at $\mu = 1/2$ naturally arises from the combination of Tikhonov regularization, that shows a well-known saturation phenomenon at $\mu = 1$, with a discrepancy principle as a stopping rule that has been proven to shift the saturation exponent (also called qualification) of linear regularization methods by $-1/2$ (cf. [45]). We want to mention that Rieder in [137] also considers the application of other regularization methods than Tikhonov regularization in each Newton step. Generalizations of this kind will be considered in Section 4.3 below. Due to the difficulty of showing convergence rates, we will not follow the Levenberg–Marquardt approach there but the method introduced in Section 4.2 below.

Using a-priori strategies instead of the a-posteriori choices (2.2) and (4.3) for k_* and α_k, respectively, we will show in Theorem 4.7 below, that almost optimal rates can be obtained under the smoothness condition (2.22) for any μ in the interval $(0, 1/2]$ provided that condition (4.12) holds. Compared to a-posteriori strategies this has the drawback that μ (or a positive lower bound on it) has to be known to guarantee convergence rates, but also the advantages that (except for strict positivity) no restriction like (4.11) on μ has to hold and that one saves the effort of computing α_k from the nonlinear equation (4.3).

If (4.12) holds, the range of the adjoint of the Fréchet-derivative, $F'(x)^*$, and hence of $(F'(x)^* F'(x))^{\frac{1}{2}}$ is locally invariant. By the inequality of Heinz, this also implies range invariance of all powers of $(F'(x)^* F'(x))$ smaller than $1/2$.

Lemma 4.4 *Let $K \in \mathcal{L}(\mathcal{X}, \mathcal{Y})$ and $R \in \mathcal{L}(\mathcal{Y}, \mathcal{Y})$ with $\|I - R\| < 1/2$. Then for any $\mu \in [0, 1/2]$ it holds that $\mathcal{R}((K^* R^* R K)^\mu) = \mathcal{R}((K^* K)^\mu)$ and that*

$$\|((K^* R^* R K)^\mu)^\dagger (K^* K)^\mu\| \;\le\; (1 - \|I - R\|)^{-2\mu}\,, \tag{4.13}$$

$$\|((K^* K)^\mu)^\dagger (K^* R^* R K)^\mu\| \;\le\; (1 - \|I - R\|)^{2\mu} (1 - 2\|I - R\|)^{-2\mu}. \tag{4.14}$$

Proof. We use the inequality of Heinz (see, e.g., [45, Proposition 8.21]), which says that for two densely defined unbounded selfadjoint strictly positive operators L and A on a Hilbert space $\mathcal{H}$ with

$$\mathcal{D}(A) \subseteq \mathcal{D}(L) \quad \text{and} \quad \|Lx\| \leq \|Ax\| \quad \text{for all } x \in \mathcal{D}(A)\,,$$

it also holds that

$$\mathcal{D}(A^{\kappa}) \subseteq \mathcal{D}(L^{\kappa}) \quad \text{and} \quad \|L^{\kappa}x\| \leq \|A^{\kappa}x\| \quad \text{for all } x \in \mathcal{D}(A^{\kappa})\,, \tag{4.15}$$

for all $\kappa \in [0,1]$. Setting $A := ((K^*K)^{\frac{1}{2}})^{\dagger}$, $L := (1 - \|I - R\|)((K^*R^*RK)^{\frac{1}{2}})^{\dagger}$, and $\mathcal{H} := \mathcal{N}((K^*K)^{\frac{1}{2}})^{\perp} = \mathcal{N}(K)^{\perp} = \mathcal{N}(RK)^{\perp} = \mathcal{N}((K^*R^*RK)^{\frac{1}{2}})^{\perp} \subseteq \mathcal{X}$, with the norm of $\mathcal{X}$, we obtain

$$\mathcal{D}(A) = \mathcal{R}((K^*K)^{\frac{1}{2}}) = \mathcal{R}(K^*) = \mathcal{R}(K^*R^*) = \mathcal{R}((K^*R^*RK)^{\frac{1}{2}}) = \mathcal{D}(L)\,.$$

Note that, since $\|I - R\| < 1$, by the Neumann series expansion, R and R^* are invertible with

$$\|R^{-1}\| \leq \frac{1}{1 - \|I - R\|} \qquad \text{and} \qquad \|I - R^{-1}\| \leq \frac{\|I - R\|}{1 - \|I - R\|}\,. \tag{4.16}$$

Since the operators

$$\begin{aligned} T &:= (K^*K)^{-\frac{1}{2}}K^* : \mathcal{N}(K^*)^{\perp} \to \mathcal{H}\,, \\ \tilde{T} &:= (K^*R^*)^{-1}(K^*R^*RK)^{\frac{1}{2}} : \mathcal{H} \to \mathcal{N}((K^*R^*)^{\perp} \end{aligned}$$

are isomorphisms, for any $x \in \mathcal{D}(A) = \mathcal{D}(L)$, i.e., $x = (K^*R^*RK)^{\frac{1}{2}}w$ for some $w \in \mathcal{H}$, we have that

$$\|Ax\| = \|TR^*\tilde{T}w\| \geq (1 - \|I - R\|)\,\|w\| = \|Lx\|\,.$$

Therefore, (4.15) with $\kappa = 2\mu$ now implies (4.13).

The identity $\mathcal{R}((K^*R^*RK)^{\mu}) = \mathcal{R}((K^*K)^{\mu})$ follows from a further application of (4.15) noting that one can show similiarly as above that also the estimate

$$\|Ax\| \leq \frac{2}{1 - \|I - R\|}\,\|Lx\|$$

holds. Finally, (4.14) follows as above by interchanging the role of K and RK using (4.16). Note that, since $\|I - R\| < 1/2$, (4.16) implies that $\|I - R^{-1}\| < 1$. □

The idea of the convergence proof given below is now that if (2.22) holds with $\mu \leq 1/2$ for the initial guess x_0, then all the subsequent iterates also remain in the range of $(F'(x^\dagger)^*F'(x^\dagger))^\mu$ and their preimages under $(F'(x^\dagger)^*F'(x^\dagger))^\mu$, $v_k^\delta := ((F'(x^\dagger)^*F'(x^\dagger))^\mu)^\dagger(x_k - x^\dagger)$, are uniformly bounded.

To prove this, we need the estimates given in the following two lemmata:

Lemma 4.5 *Let $K \in \mathcal{L}(\mathcal{X},\mathcal{Y})$, $s \in [0,1]$, and let $\{\alpha_k\}$ be a sequence satisfying $\alpha_k > 0$ and $\alpha_k \to 0$ as $k \to \infty$. Then it holds that*

$$w_k(s) := \alpha_k^{1-s}\,\|(K^*K+\alpha_k I)^{-1}(K^*K)^s v\| \leq s^s(1-s)^{1-s}\,\|v\| \leq \|v\| \quad (4.17)$$

and that

$$\lim_{k\to\infty} w_k(s) = \begin{cases} 0\,, & 0 \leq s < 1\,, \\ \|v\|\,, & s = 1\,, \end{cases} \quad (4.18)$$

for any $v \in \mathcal{N}(A)^\perp$.

Proof. The assertions follow with spectral theory similar as in the proof of Lemma 2.10. □

Lemma 4.6 *Let* (4.12) *hold and assume that $x_k^\delta \in \mathcal{B}_\rho(x_0)$. Moreover, we put $K := F'(x^\dagger)$ and $e_k^\delta := x_k^\delta - x^\dagger$ and assume that $e_k^\delta = (K^*K)^\mu v_k^\delta$ for some $0 < \mu \leq 1/2$ and some $v_k^\delta \in \mathcal{X}$ and that $c_R\|e_k^\delta\| < 1/2$. Then there exists a $v_{k+1}^\delta \in \mathcal{X}$ such that $e_{k+1}^\delta = (K^*K)^\mu v_{k+1}^\delta$ and the following estimates hold:*

$$\begin{aligned} \|v_{k+1}^\delta\| \;&\leq\; (1-2c_R\|e_k^\delta\|)^{-2\mu}\,\|v_k^\delta\| + (1-c_R\|e_k^\delta\|)^{2\mu} \\ &\qquad (1-2c_R\|e_k^\delta\|)^{-2\mu}\alpha_k^{-(\mu+\frac{1}{2})}(\tfrac{3}{2}c_R\|e_k^\delta\|\,\|Ke_k^\delta\|+\delta)\,, \end{aligned} \quad (4.19)$$

$$\begin{aligned} \|e_{k+1}^\delta\| \;&\leq\; \alpha_k^\mu(1-c_R\|e_k^\delta\|)^{-2\mu}\,\|v_k^\delta\| \\ &\qquad + \tfrac{1}{2}\alpha_k^{-\frac{1}{2}}(\tfrac{3}{2}c_R\|e_k^\delta\|\,\|Ke_k^\delta\|+\delta)\,, \end{aligned} \quad (4.20)$$

$$\begin{aligned} \|Ke_{k+1}^\delta\| \;&\leq\; (1-c_R\|e_k^\delta\|)^{-1}\Big((1-c_R\|e_k^\delta\|)^{-2\mu}\alpha_k^{\mu+\frac{1}{2}}\,\|v_k^\delta\| \\ &\qquad + \tfrac{3}{2}c_R\|e_k^\delta\|\,\|Ke_k^\delta\|+\delta\Big)\,. \end{aligned} \quad (4.21)$$

Proof. Denoting $K_k := F'(x_k^\delta)$, we can rewrite the error as follows,

$$\begin{aligned} e_{k+1}^\delta \;&=\; \alpha_k(K_k^*K_k+\alpha_k I)^{-1}(K^*K)^\mu v_k^\delta \\ &\qquad + (K_k^*K_k+\alpha_k I)^{-1}K_k^*(y^\delta - F(x_k^\delta)+K_k e_k^\delta)\,. \end{aligned} \quad (4.22)$$

This, due to (4.12), (4.13), Lemma 4.5, and the Taylor remainder estimate

$$\begin{aligned} &\|F(x^\dagger) - F(x) + F'(x)(x - x^\dagger)\| \\ &\quad = \left\| \int_0^1 (F'(x^\dagger + t(x - x^\dagger)) - F'(x))(x - x^\dagger)\, dt \right\| \\ &\quad = \left\| \int_0^1 (R_{x^\dagger + t(x - x^\dagger)} - R_x) F'(x^\dagger)(x - x^\dagger)\, dt \right\| \\ &\quad \le \tfrac{3}{2} c_R \|x - x^\dagger\| \, \|F'(x^\dagger)(x - x^\dagger)\| \end{aligned} \tag{4.23}$$

for $x \in \mathcal{B}_\rho(x_0)$, as well as (1.2), implies (4.20). Note that

$$\|I - R_{x_k^\delta}\| \le c_R \|e_k^\delta\| < \tfrac{1}{2}\,. \tag{4.24}$$

By Lemma 4.4 and (4.12), the error representation (4.22) also implies that e_{k+1}^δ is in the range of $(K^*K)^\mu$. Its preimage under $(K^*K)^\mu$ can be represented as

$$\begin{aligned} v_{k+1}^\delta \;=\; & ((K^*K)^\mu)^\dagger (K_k^*K_k)^\mu \Big(\alpha_k (K_k^*K_k + \alpha_k I)^{-1} ((K_k^*K_k)^\mu)^\dagger (K^*K)^\mu v_k^\delta \\ & + (K_k^*K_k + \alpha_k I)^{-1} ((K_k^*K_k)^\mu)^\dagger K_k^* (y^\delta - F(x_k^\delta) + K_k e_k^\delta) \Big). \end{aligned}$$

This together with Lemma 4.5, (1.2), (4.12), (4.13), (4.14), (4.23), and (4.24) yields (4.19).

Finally, to derive (4.21), we apply K on both sides of (4.22) and use (4.12) to obtain

$$\begin{aligned} K e_{k+1}^\delta \;=\; & R_{x_k^\delta}^{-1} \Big(\alpha_k K_k (K_k^*K_k + \alpha_k I)^{-1} (K^*K)^\mu v_k^\delta \\ & + K_k (K_k^*K_k + \alpha_k I)^{-1} K_k^* (y^\delta - F(x_k^\delta) + K_k e_k^\delta) \Big). \end{aligned}$$

Now (4.21) again follows with Lemma 4.5, (1.2), (4.12), (4.13), (4.16), (4.23), and (4.24). □

In order to deduce convergence rates from these estimates, here both α_k and the stopping index k_* are chosen a-priori, namely according to

$$\alpha_k = \alpha_0 q^k\,, \qquad \text{for some} \quad \alpha_0 > 0\,, \quad q \in (0,1)\,, \tag{4.25}$$

and

$$\begin{aligned} &\eta_{k_*} \alpha_{k_*}^{\mu + \frac{1}{2}} \le \delta < \eta_k \alpha_k^{\mu + \frac{1}{2}}\,, \qquad 0 \le k < k_*\,, \\ &\eta_k := \eta (k+1)^{-(1+\varepsilon)}\,, \qquad \text{for some} \quad \eta > 0\,, \quad \varepsilon > 0\,. \end{aligned} \tag{4.26}$$

Theorem 4.7 *Let a solution $x^\dagger$ of* (1.1) *exist and let* (4.12) *as well as* (2.22) *hold with some $0 < \mu \leq 1/2$ and $\|v\|$ sufficiently small. Moreover, let α_k and k_* be chosen according to* (4.25) *and* (4.26) *with η sufficiently small, respectively. Then the Levenberg–Marquardt iterates defined by* (4.2) *remain in $\mathcal{B}_\rho(x_0)$ and converge with the rate*

$$\|x_{k_*}^\delta - x^\dagger\| = O\Big(\big(\delta\,(1 + |\ln\delta|)^{(1+\varepsilon)}\big)^{\frac{2\mu}{2\mu+1}}\Big).$$

Moreover,

$$\|F(x_{k_*}^\delta) - y\| = O\Big(\delta\,(1 + |\ln\delta|)^{(1+\varepsilon)}\Big)$$

and

$$k_* = O(1 + |\ln\delta|)\,.$$

For the noise free case ($\delta = 0$, $\eta = 0$) we obtain that

$$\|x_k - x^\dagger\| = O(\alpha_k^\mu)\,,$$

and that

$$\|F(x_k) - y\| = O\Big(\alpha_k^{\mu+\frac{1}{2}}\Big).$$

Proof. Let $\tilde{\gamma} > 0$, $\|v\|$, and η satisfy the following smallness assumptions,

$$q^{-\mu}\tilde{\gamma}\alpha_0^\mu \leq \min\{\tfrac{1}{3c_R}\,,\,\tfrac{\rho}{2}\}\,, \tag{4.27}$$

$$\|v\| \leq q^{-\mu}\tilde{\gamma}\alpha_0^\mu\,\|K\|^{-2\mu}\min\{1\,,\,\tfrac{3}{2}q^{-\frac{1}{2}}\alpha_0^{\frac{1}{2}}\,\|K\|^{-1}\}\,, \tag{4.28}$$

and

$$\exp\left(\frac{12\mu c_R\tilde{\gamma}\alpha_0^\mu}{q^\mu(1-q^\mu)}\right)\left(\|v\| + \frac{9c_R\tilde{\gamma}^2\alpha_0^\mu}{4q^{(2\mu+\frac{1}{2})}(1-q^\mu)} + \eta c_\varepsilon\right) \leq \tilde{\gamma}\,, \tag{4.29}$$

where c_ε is defined by

$$c_\varepsilon := \sum_{j=0}^{\infty}(j+1)^{-(1+\varepsilon)}. \tag{4.30}$$

Note that this can always be achieved, e.g., by choosing first $\tilde{\gamma} > 0$ so small that (4.27) is satisfied and that

$$\exp\left(\frac{12\mu c_R\tilde{\gamma}\alpha_0^\mu}{q^\mu(1-q^\mu)}\right) \leq 2 \qquad \text{and} \qquad \frac{9c_R\tilde{\gamma}^2\alpha_0^\mu}{4q^{(2\mu+\frac{1}{2})}(1-q^\mu)} < \frac{\tilde{\gamma}}{6}\,.$$

Then one chooses $\|v\|$ so small that (4.28) holds and that $\|v\| \leq \tilde{\gamma}/6$. Finally, one chooses η so small that $\eta c_\varepsilon < \tilde{\gamma}/6$. This yields (4.29).

Moreover, let $\{\gamma_k^\delta\}$ be a sequence defined recursively via

$$\begin{aligned}
\gamma_{k+1} &:= (1 + 12\mu c_R q^{-\mu}\tilde{\gamma}\alpha_k^\mu)\gamma_k \\
&\qquad + (1 + 6\mu c_R q^{-\mu}\tilde{\gamma}\alpha_k^\mu)(\tfrac{9}{4}c_R q^{-(2\mu+\frac{1}{2})}\tilde{\gamma}^2\alpha_k^\mu + \eta_k), \qquad (4.31)\\
\gamma_0 &:= \|v\|.
\end{aligned}$$

With the notation of Lemma 4.6, we will show by induction that

$$\begin{aligned}
e_k^\delta &= (K^*K)^\mu v_k^\delta, & \|v_k^\delta\| &\le \gamma_k, \\
\|e_k^\delta\| &\le q^{-\mu}\tilde{\gamma}\alpha_k^\mu \le \tfrac{\rho}{2}, & \|Ke_k^\delta\| &\le \tfrac{3}{2}q^{-(\mu+\frac{1}{2})}\tilde{\gamma}\alpha_k^{\mu+\frac{1}{2}}
\end{aligned} \qquad (4.32)$$

hold for all $k \le k_*$. Note that $\|e_k^\delta\| \le \rho/2$ guarantees that x_k^δ remains in $\mathcal{B}_\rho(x_0)$.

For $k = 0$, (4.32) follows directly from (2.22), $v_0 := v$, $\gamma_0 = \|v\|$, (4.27), (4.28), and the estimates $\|e_0\| \le \|K\|^{2\mu}\|v\|$ as well as $\|Ke_0\| \le \|K\|^{2\mu+1}\|v\|$.

Assuming that (4.32) holds for some $0 < k < k_*$, we now apply Lemma 4.6. Together with (4.25), (4.26), (4.31), and the simple estimates

$$\begin{aligned}
&\left(\frac{1}{1-s}\right)^{2\mu} \le 1 + \frac{2\mu s}{1-s} \le 1 + 3\mu s, \\
&\left(\frac{1}{1-2s}\right)^{2\mu} \le 1 + \frac{4\mu s}{1-2s} \le 1 + 12\mu s, \\
&\left(\frac{1-s}{1-2s}\right)^{2\mu} \le 1 + \frac{2\mu s}{1-2s} \le 1 + 6\mu s, \qquad \frac{1}{1-s} \le \frac{3}{2},
\end{aligned}$$

that follow from the Bernoulli inequality provided that $0 \le s \le 1/3$ and $\mu \le 1/2$, we obtain the estimates

$$\begin{aligned}
\|v_{k+1}^\delta\| &\le (1 + 12\mu c_R q^{-\mu}\tilde{\gamma}\alpha_k^\mu)\gamma_k \\
&\qquad + (1 + 6\mu c_R q^{-\mu}\tilde{\gamma}\alpha_k^\mu)(\tfrac{9}{4}c_R q^{-(2\mu+\frac{1}{2})}\tilde{\gamma}^2\alpha_k^\mu + \eta_k) = \gamma_{k+1}, \\
\|e_{k+1}^\delta\| &\le \alpha_k^\mu[(1 + 3\mu c_R q^{-\mu}\tilde{\gamma}\alpha_k^\mu)\gamma_k + \tfrac{1}{2}(\tfrac{9}{4}c_R q^{-(2\mu+\frac{1}{2})}\tilde{\gamma}^2\alpha_k^\mu + \eta_k)] \\
&\le q^{-\mu}\alpha_{k+1}^\mu\gamma_{k+1}, \\
\|Ke_{k+1}^\delta\| &\le \tfrac{3}{2}\alpha_k^{\mu+\frac{1}{2}}[(1 + 3\mu c_R q^{-\mu}\tilde{\gamma}\alpha_k^\mu)\gamma_k + (\tfrac{9}{4}c_R q^{-2(\mu+\frac{1}{2})}\tilde{\gamma}^2\alpha_k^\mu + \eta_k)] \\
&\le \tfrac{3}{2}q^{-\mu+\frac{1}{2}}\alpha_{k+1}^{\mu+\frac{1}{2}}\gamma_{k+1}.
\end{aligned}$$

Note that, due to (4.25) and (4.27), $c_R q^{-\mu}\tilde{\gamma}\alpha_k^\mu \le 1/3$. Thus, (4.32) holds for all $k \le k_*$ if we can show that

$$\gamma_k^\delta \le \tilde{\gamma} \qquad \text{for all } k \in \mathbb{N}. \qquad (4.33)$$

From the recursion formula (4.31) one immediately gets the representation

$$\begin{aligned}\gamma_k &= \Big(\prod_{j=0}^{k-1}(1+12\mu c_R q^{-\mu}\tilde{\gamma}\alpha_j^\mu)\Big)\gamma_0 + \sum_{j=0}^{k-1}\Big(\prod_{l=j+1}^{k-1}(1+12\mu c_R q^{-\mu}\tilde{\gamma}\alpha_l^\mu)\Big)\\ &\qquad \cdot(1+6\mu c_R q^{-\mu}\tilde{\gamma}\alpha_j^\mu)(\tfrac{9}{4}c_R q^{-(2\mu+\frac{1}{2})}\tilde{\gamma}^2\alpha_j^\mu + \eta_j)\end{aligned}$$

which together with (4.25), (4.26), (4.29), (4.30), $\gamma_0 = \|v\|$, and the estimate

$$\prod_{j=0}^{k-1}(1+a_j) \le \exp\Big(\sum_{j=0}^{k-1} a_j\Big), \qquad a_j \ge 0\,, \quad 0 \le j \le k-1\,,$$

yields

$$\begin{aligned}\gamma_k &\le \Big(\prod_{j=0}^{k-1}(1+12\mu c_R q^{-\mu}\tilde{\gamma}\alpha_j^\mu)\Big)\Big(\gamma_0 + \sum_{j=0}^{k-1}(\tfrac{9}{4}c_R q^{-(2\mu+\frac{1}{2})}\tilde{\gamma}^2\alpha_j^\mu + \eta_j)\Big)\\ &\le \exp\Big(12\mu c_R q^{-\mu}\tilde{\gamma}\sum_{j=0}^{k-1}\alpha_j^\mu\Big)\Big(\gamma_0 + \sum_{j=0}^{k-1}(\tfrac{9}{4}c_R q^{-(2\mu+\frac{1}{2})}\tilde{\gamma}^2\alpha_j^\mu + \eta_j)\Big)\\ &\le \exp\left(\frac{12\mu c_R\tilde{\gamma}\alpha_0^\mu}{q^\mu(1-q^\mu)}\right)\left(\gamma_0 + \frac{9c_R\tilde{\gamma}^2\alpha_0^\mu}{4q^{(2\mu+\frac{1}{2})}(1-q^\mu)} + \eta c_\varepsilon\right) \le \tilde{\gamma}\,.\end{aligned}$$

Thus, (4.33) is shown.

Since we obtain analogoulsy to (4.23) that

$$\|F(x_k^\delta) - F(x^\dagger) - Ke_k^\delta\| \le \tfrac{1}{2}c_R\|e_k^\delta\|\,\|Ke_k^\delta\|\,,$$

(4.27) and (4.32) imply that

$$\|F(x_k^\delta) - y\| \le \tfrac{7}{4}q^{-(\mu+\frac{1}{2})}\tilde{\gamma}\alpha_k^{\mu+\frac{1}{2}}. \tag{4.34}$$

This together with (4.32) implies the assertions of the theorem for the noise free case $\delta = 0$, $\eta = 0$.

In case of noisy data, (4.32) with $k = k_*$, as well as the stopping rule (4.26) yield

$$\|e_{k_*}^\delta\| = O(\alpha_{k_*}^\mu) = O\Big((\delta\eta_{k_*}^{-1})^{\frac{2\mu}{2\mu+1}}\Big). \tag{4.35}$$

Now (4.25) and (4.26) imply that

$$\delta \le \eta_k\alpha_k^{\mu+\frac{1}{2}} \le \eta\alpha_0^{\mu+\frac{1}{2}}q^{(\mu+\frac{1}{2})k}$$

for $k < k_*$ and hence that

$$k_* = O(1 + |\ln\delta|) .$$

This together with (4.26), (4.34), and (4.35) proves the assertions for the noisy case. □

4.2 Iteratively regularized Gauss–Newton method

In this section we deal with the iteratively regularized Gauss–Newton method

$$x_{k+1}^\delta = x_k^\delta + (F'(x_k^\delta)^* F'(x_k^\delta) + \alpha_k I)^{-1}(F'(x_k^\delta)^*(y^\delta - F(x_k^\delta)) + \alpha_k(x_0 - x_k^\delta)), \tag{4.36}$$

where, as always, $x_0^\delta = x_0$ is an initial guess for the true solution, α_k is a sequence of positive numbers tending towards zero, and y^δ are noisy data satisfying the estimate (1.2). This method is quite similar to Levenberg–Marquardt iterations (4.2).

Note that the approximate solution x_{k+1}^δ minimizes the functional

$$\phi(x) := \|y^\delta - F(x_k^\delta) - F'(x_k^\delta)(x - x_k^\delta)\|^2 + \alpha_k \|x - x_0\|^2.$$

This means that x_{k+1}^δ minimizes the Tikhonov functional where the nonlinear function F is linearized around x_k^δ (cf. [45, Chapter 10]).

We emphasize that for a fixed number of iterations the process (4.36) is a stable algorithm if $F'(\cdot)$ is continuous.

This method was introduced by Bakushinskii. In [3] he proved local convergence essentially under the source condition (2.22) with $\mu \geq 1$ assuming that F' is Lipschitz continuous. For the noise free case he even proved the rate

$$\|x_n - x^\dagger\| = O(\alpha_k) .$$

In [11] it was shown that convergence rates can even be obtained if $\mu < 1$ and also for the noisy case. A convergence analysis for a continuous analogue of the iteratively regularized Gauss–Newton method was derived in [92].

As for Landweber iteration (cf. Chapter 2), Lipschitz continuity of F' is not sufficient to obtain rates if $\mu < 1/2$ in (2.22). Similarly to Assumption 2.7 (ii) we then need further conditions on F' that guarantee that the linearization is not too far away from the nonlinear operator F. However, for the case $\mu \geq 1/2$, Lipschitz continuity of F' suffices to prove convergence rates for the iteratively regularized Gauss–Newton method as for Tikhonov regularization or the iteratively regularized Landweber iteration.

For our convergence analysis we need an assumption similar to Assumption 2.7.

Assumption 4.8 Let ρ be a positive number such that $\mathcal{B}_{2\rho}(x_0) \subset \mathcal{D}(F)$.

(i) The equation $F(x) = y$ has an x_0-minimum-norm solution $x^\dagger$ in $\mathcal{B}_\rho(x_0)$.

(ii) $x^\dagger$ satisfies the source condition (2.22) for some $0 \le \mu \le 1$, i.e.,

$$x^\dagger - x_0 = (F'(x^\dagger)^* F'(x^\dagger))^\mu v\,, \quad v \in \mathcal{N}(F'(x^\dagger))^\perp.$$

(iii) If $\mu < 1/2$, the Fréchet-derivative F' satisfies the conditions

$$F'(\tilde{x}) = R(\tilde{x}, x)F'(x) + Q(\tilde{x}, x) \tag{4.37}$$

$$\|I - R(\tilde{x}, x)\| \le c_R \tag{4.38}$$

$$\|Q(\tilde{x}, x)\| \le c_Q \|F'(x^\dagger)(\tilde{x} - x)\| \tag{4.39}$$

for $x, \tilde{x} \in \mathcal{B}_{2\rho}(x_0)$, where c_R and c_Q are nonnegative constants.

If $\mu \ge 1/2$, the Fréchet-derivative F' is Lipschitz continuous in $\mathcal{B}_{2\rho}(x_0)$, i.e.,

$$\|F'(\tilde{x}) - F'(x)\| \le L \|\tilde{x} - x\|\,, \qquad x, \tilde{x} \in \mathcal{B}_{2\rho}(x_0) \tag{4.40}$$

for some $L > 0$.

(iv) The sequence $\{\alpha_k\}$ in (4.36) satisfies

$$\alpha_k > 0\,, \qquad 1 \le \frac{\alpha_k}{\alpha_{k+1}} \le r\,, \qquad \lim_{k\to\infty} \alpha_k = 0\,, \tag{4.41}$$

for some $r > 1$.

We show in the following proposition that the conditions (4.37) – (4.39) imply that F is constant on $x^\dagger + \mathcal{N}(F'(x^\dagger)) \cap \mathcal{B}_\rho(x^\dagger)$ assuming that ρ, c_R and c_Q are sufficiently small (compare (2.5)). Note that these conditions are slightly weaker than (2.26) and (2.27) for Landweber iterates. Moreover, we prove that the condition $x^\dagger - x_0 \in \mathcal{N}(F'(x^\dagger))^\perp$, which is an immediate consequence of Assumption 4.8 (ii), is not restrictive. It is automatically satisfied for the then unique x_0-minimum-norm solution (compare Proposition 2.1).

Proposition 4.9 *Let $x^\dagger \in \mathcal{B}_\rho(x_0) \subset \mathcal{D}(F)$ be a solution of $F(x) = y$ and let the conditions* (4.37) – (4.39) *hold.*

(i) If $x - x^\dagger \in \mathcal{N}(F'(x^\dagger))$ and $x \in \mathcal{B}_\rho(x_0)$, then $F(x) = F(x^\dagger) = y$. Moreover, if $c_R + \rho c_Q < 1$, then

$$x - x^\dagger \in \mathcal{N}(F'(x^\dagger)) \iff F(x) = F(x^\dagger)$$

holds in $\mathcal{B}_\rho(x_0)$.

(ii) If $x^\dagger - x_0 \in \mathcal{N}(F'(x^\dagger))^\perp$ and if $c_R + \rho c_Q < 1$, then $x^\dagger$ is the unique x_0-minimum-norm solution in $\mathcal{B}_\rho(x_0)$. On the other hand, if $x^\dagger$ is an x_0-minimum-norm solution in $\mathcal{B}_\rho(x_0)$, then $x^\dagger - x_0 \in \mathcal{N}(F'(x^\dagger))^\perp$.

Proof. (i) Let $x \in \mathcal{B}_\rho(x_0)$. Then conditions (4.37) – (4.39) imply the following estimate

$$\begin{aligned}
&\|F(x) - F(x^\dagger) - F'(x^\dagger)(x - x^\dagger)\| \\
&\quad = \Big\| \int_0^1 (F'(x^\dagger + t(x - x^\dagger)) - F'(x^\dagger))(x - x^\dagger)\, dt \Big\| \\
&\quad = \Big\| \int_0^1 [\,(R(x^\dagger + t(x - x^\dagger), x^\dagger) - I)F'(x^\dagger) \\
&\qquad\qquad + Q(x^\dagger + t(x - x^\dagger), x^\dagger)\,](x - x^\dagger)\, dt \Big\| \\
&\quad \le (c_R + \tfrac{1}{2} c_Q \|x - x^\dagger\|)\, \|F'(x^\dagger)(x - x^\dagger)\|
\end{aligned}$$

and hence

$$\|F(x) - F(x^\dagger)\| \le (1 + c_R + \tfrac{1}{2} c_Q \|x - x^\dagger\|)\, \|F'(x^\dagger)(x - x^\dagger)\| \tag{4.42}$$

and

$$\|F(x) - F(x^\dagger)\| \ge (1 - c_R - \tfrac{1}{2} c_Q \|x - x^\dagger\|)\, \|F'(x^\dagger)(x - x^\dagger)\|. \tag{4.43}$$

Since $\|x - x^\dagger\| \le 2\rho$, both estimates imply the assertions.
The proof of the assertions in (ii) is similar to the one of Proposition 2.1. □

Before we can prove convergence or convergence rates for the iteration process (4.36) we need some preparatory lemmata. In the first lemma we derive an estimate for $\|e_{k+1}^\delta\|$ and $\|Ke_{k+1}^\delta\|$ assuming that $x_k^\delta \in \mathcal{B}_\rho(x^\dagger)$.

Lemma 4.10 *Let Assumption 4.8 hold and assume that $x_k^\delta \in \mathcal{B}_\rho(x^\dagger)$. Moreover, set $K := F'(x^\dagger)$, $e_k^\delta := x_k^\delta - x^\dagger$, and let $w_k(\cdot)$ be defined as in* (4.17).

(i) If $0 \le \mu < 1/2$, we obtain the estimates

$$\begin{aligned}
\|e_{k+1}^\delta\| \le\; & \alpha_k^\mu w_k(\mu) + c_R \alpha_k^\mu w_k(\mu + \tfrac{1}{2}) \\
& + c_Q \|Ke_k^\delta\|\, \alpha_k^{\mu - \frac{1}{2}} (\tfrac{1}{2} w_k(\mu) + w_k(\mu + \tfrac{1}{2})) \\
& + \alpha_k^{-\frac{1}{2}} (c_R \|Ke_k^\delta\| + \tfrac{3}{4} c_Q \|e_k^\delta\|\, \|Ke_k^\delta\| + \tfrac{1}{2}\delta),
\end{aligned} \tag{4.44}$$

$$\begin{aligned}
\|Ke_{k+1}^\delta\| \le\ & (1+2c_R(1+c_R))\alpha_k^{\mu+\frac{1}{2}} w_k(\mu+\tfrac{1}{2}) \\
& + \|Ke_k^\delta\| \Big(c_Q(1+c_R)\alpha_k^\mu (w_k(\mu) + \tfrac{1}{2} w_k(\mu+\tfrac{1}{2})) \\
& \qquad + c_Q c_R \alpha_k^\mu w_k(\mu+\tfrac{1}{2}) \\
& \qquad + (1+c_R)(2c_R + \tfrac{3}{2} c_Q \|e_k^\delta\|) \Big) \qquad (4.45) \\
& + \|Ke_k^\delta\|^2 c_Q \Big(c_Q \alpha_k^{\mu-\frac{1}{2}} (\tfrac{1}{2} w_k(\mu) + w_k(\mu+\tfrac{1}{2})) \\
& \qquad + \alpha_k^{-\frac{1}{2}} (c_R + \tfrac{3}{4} c_Q \|e_k^\delta\|) \Big) \\
& + (1 + c_R + \tfrac{1}{2} c_Q \alpha_k^{-\frac{1}{2}} \|Ke_k^\delta\|)\, \delta\,.
\end{aligned}$$

(ii) If $1/2 \le \mu \le 1$, we obtain the estimates

$$\begin{aligned}
\|e_{k+1}^\delta\| \le\ & \alpha_k^\mu w_k(\mu) + L \|e_k^\delta\| \Big(\tfrac{1}{2} \alpha_k^{\mu-\frac{1}{2}} w_k(\mu) \qquad (4.46) \\
& + \|(K^*K)^{\mu-\frac{1}{2}} v\| \Big) + \tfrac{1}{2} \alpha_k^{-\frac{1}{2}} (\tfrac{1}{2} L \|e_k^\delta\|^2 + \delta)\,, \\
\|Ke_{k+1}^\delta\| \le\ & \alpha_k \|(K^*K)^{\mu-\frac{1}{2}} v\| \\
& + L^2 \|e_k^\delta\|^2 (\tfrac{1}{2} \alpha_k^{\mu-\frac{1}{2}} w_k(\mu) + \|(K^*K)^{\mu-\frac{1}{2}} v\|) \qquad (4.47) \\
& + L \alpha_k^{\frac{1}{2}} \|e_k^\delta\| (\alpha_k^{\mu-\frac{1}{2}} w_k(\mu) + \tfrac{1}{2} \|(K^*K)^{\mu-\frac{1}{2}} v\|) \\
& + (\tfrac{1}{2} L \alpha_k^{-\frac{1}{2}} \|e_k^\delta\| + 1)(\tfrac{1}{2} L \|e_k^\delta\|^2 + \delta)\,.
\end{aligned}$$

Proof. We set $K_k := F'(x_k^\delta)$. Due to (2.22), we can rewrite (4.36) as follows

$$\begin{aligned}
e_{k+1}^\delta =\ & -\alpha_k (K^*K + \alpha_k I)^{-1} (K^*K)^\mu v \\
& - \alpha_k (K_k^* K_k + \alpha_k I)^{-1} \Big(K_k^* (K - K_k) \qquad (4.48) \\
& \qquad + (K^* - K_k^*) K \Big) (K^*K + \alpha_k I)^{-1} (K^*K)^\mu v \\
& + (K_k^* K_k + \alpha_k I)^{-1} K_k^* (y^\delta - F(x_k^\delta) + K_k e_k^\delta)\,.
\end{aligned}$$

(i) Let us first consider the case that $0 \le \mu < 1/2$. Since $x_k^\delta \in \mathcal{B}_\rho(x^\dagger) \subset \mathcal{B}_{2\rho}(x_0)$, the conditions (4.37) – (4.39) are applicable. By reasoning similar to that in the proof of Proposition 4.9, we obtain that

$$\|F(x_k^\delta) - F(x^\dagger) - K_k e_k^\delta\| \le (2c_R + \tfrac{3}{2} c_Q \|e_k^\delta\|) \|Ke_k^\delta\|. \qquad (4.49)$$

The well-known estimates (see (4.17))

$$\|(K_k^*K_k + \alpha_k I)^{-1}\| \le \alpha_k^{-1}, \qquad \|(K_k^*K_k + \alpha_k I)^{-1}K_k^*\| \le \tfrac{1}{2}\alpha_k^{-\frac{1}{2}},$$

and

$$\begin{aligned} K_k^*(K - K_k) + (K^* - K_k^*)K &= K_k^*(R^*(x^\dagger, x_k^\delta) - R(x_k^\delta, x^\dagger))K \\ &\quad + Q^*(x^\dagger, x_k^\delta)K - K_k^*Q(x_k^\delta, x^\dagger) \end{aligned}$$

imply that

$$\begin{aligned} &\|\alpha_k(K_k^*K_k + \alpha_k I)^{-1}(K_k^*(K - K_k) \\ &\qquad + (K^* - K_k^*)K)(K^*K + \alpha_k I)^{-1}(K^*K)^\mu v\| \\ &\qquad\le \Big(\tfrac{1}{2}\alpha_k^{\frac{1}{2}}\|R^*(x^\dagger, x_k^\delta) - R(x_k^\delta, x^\dagger)\| + \|Q^*(x^\dagger, x_k^\delta)\|\Big) \\ &\qquad\quad \cdot \|K(K^*K + \alpha_k I)^{-1}(K^*K)^\mu v\| \\ &\qquad\quad + \tfrac{1}{2}\alpha_k^{\frac{1}{2}}\|Q(x_k^\delta, x^\dagger)\|\,\|(K^*K + \alpha_k I)^{-1}(K^*K)^\mu v\|. \end{aligned}$$

This together with (1.2), (4.17), (4.48), and $F(x^\dagger) = y$ yields the estimate (4.44). Note that

$$\|K(K^*K + \alpha_k I)^{-1}(K^*K)^\mu v\| = \alpha_k^{\mu - \frac{1}{2}} w_k(\mu + \tfrac{1}{2}).$$

Since, due to (4.37), $K = R(x^\dagger, x_k^\delta)K_k + Q(x^\dagger, x_k^\delta)$, we obtain together with (4.48) that

$$\begin{aligned} Ke_{k+1}^\delta &= -\alpha_k K(K^*K + \alpha_k I)^{-1}(K^*K)^\mu v \\ &\quad - \alpha_k(R(x^\dagger, x_k^\delta)K_k + Q(x^\dagger, x_k^\delta))(K_k^*K_k + \alpha_k I)^{-1} \\ &\qquad \Big(K_k^*(R^*(x^\dagger, x_k^\delta) - R(x_k^\delta, x^\dagger))K + Q^*(x^\dagger, x_k^\delta)K \\ &\qquad\quad - K_k^*Q(x_k^\delta, x^\dagger)\Big)(K^*K + \alpha_k I)^{-1}(K^*K)^\mu v \\ &\quad - (R(x^\dagger, x_k^\delta)K_k + Q(x^\dagger, x_k^\delta))(K_k^*K_k + \alpha_k I)^{-1}K_k^* \\ &\qquad (F(x_k^\delta) - F(x^\dagger) - K_k e_k^\delta + y - y^\delta). \end{aligned}$$

Now the estimate (4.45) for $\|Ke_{k+1}^\delta\|$ follows together with (1.2), (4.38), (4.39), (4.17) and (4.49).

(ii) Let us now consider the case that $1/2 \le \mu \le 1$. Then the following estimates

$$\|K - K_k\| \le L\|e_k^\delta\| \qquad \text{and} \qquad \|F(x_k^\delta) - F(x^\dagger) - K_k e_k^\delta\| \le \tfrac{1}{2}L\|e_k^\delta\|^2$$

are valid. This together with $\|K(K^*K + \alpha_k I)^{-1}(K^*K)^{\frac{1}{2}}\| \leq 1$, (1.2), (4.17), and (4.48) implies (4.46) and (4.47). □

In the next lemma we show a general asymptotic result on sequences defined by quadratic recursions.

Lemma 4.11 *Let $\{a_k^\delta\}_{k\in\mathbb{N}_0}$, $\delta \geq 0$, be a family of sequences satisfying*

$$0 \leq a_k^\delta \leq a \qquad \text{and} \qquad \limsup_{\delta\to 0,\, k\to\infty} a_k^\delta \leq a^0$$

for some $a, a^0 \in \mathbb{R}_0^+$. Moreover, let γ_k^δ be such that

$$0 \leq \gamma_{k+1}^\delta \leq a_k^\delta + b\gamma_k^\delta + c(\gamma_k^\delta)^2\,, \qquad 0 \leq k < k_*(\delta)\,, \qquad \gamma_0^\delta := \gamma_0 \tag{4.50}$$

holds, where b, c, γ_0 are nonnegative constants, $k_*(\delta) \in \mathbb{N}_0$ for any $\delta > 0$, and $k_*(\delta) \to k_*(0) := \infty$ as $\delta \to 0$.

(i) If $c > 0$, $b + 2\sqrt{ac} \leq 1$, and $\gamma_0 \leq \overline{\gamma}(a)$, then

$$\gamma_k^\delta \leq \max\{\gamma_0, \underline{\gamma}(a)\}\,, \qquad 0 \leq k \leq k_*(\delta)\,.$$

If in addition $a^0 < a$, then

$$\limsup_{\delta\to 0} \gamma_{k_*(\delta)}^\delta \leq \underline{\gamma}(a^0)\,, \qquad \limsup_{k\to\infty} \gamma_k^0 \leq \underline{\gamma}(a^0)\,.$$

Here $\underline{\gamma}(p)$ and $\overline{\gamma}(p)$, where $p \in [0, a]$, denote the fixed points of the equation $p + b\gamma + c\gamma^2 = \gamma$, i.e.,

$$\begin{aligned} \underline{\gamma}(p) &:= \frac{2p}{1 - b + \sqrt{(1-b)^2 - 4pc}}\,, \\ \overline{\gamma}(p) &:= \frac{1 - b + \sqrt{(1-b)^2 - 4pc}}{2c}\,. \end{aligned} \tag{4.51}$$

(ii) If $c = 0$ and $b < 1$, then

$$\gamma_k^\delta \leq \gamma_0 + \frac{a}{1-b}\,, \qquad 0 \leq k \leq k_*(\delta)\,,$$

and

$$\limsup_{\delta\to 0} \gamma_{k_*(\delta)}^\delta \leq \frac{a^0}{1-b}\,, \qquad \limsup_{k\to\infty} \gamma_k^0 \leq \frac{a^0}{1-b}\,.$$

Proof. Let us first assume that $c > 0$, $b + 2\sqrt{ac} \le 1$, and $\gamma_0 \le \overline{\gamma}(a)$, i.e., we consider case (i). Since $\gamma_{k+1}^\delta \le a + b\gamma_k^\delta + c(\gamma_k^\delta)^2$ and since

$$p + b\gamma + c\gamma^2 = \gamma + c(\gamma - \underline{\gamma}(p))(\gamma - \overline{\gamma}(p)), \qquad p \in [0, a], \tag{4.52}$$

it follows by induction that

$$\gamma_k^\delta \le \max\{\gamma_0, \underline{\gamma}(a)\} \le \overline{\gamma}(a), \qquad 0 \le k \le k_*(\delta).$$

Let us now assume that $a^0 < a$ and that $0 < \varepsilon < a - a^0$ is arbitrary, but fixed. Then there is a $\tilde{\delta}(\varepsilon) > 0$ and a $\tilde{k}(\varepsilon) \in \mathbb{N}_0$ such that

$$\forall 0 \le \delta \le \tilde{\delta}(\varepsilon),\, k \ge \tilde{k}(\varepsilon) \,:\, a_k^\delta \le a^0 + \varepsilon.$$

We define the sequence $\{\tilde{a}_k^\varepsilon\}_{k\in\mathbb{N}_0}$ via

$$\tilde{a}_k^\varepsilon := \begin{cases} \sup\limits_{0\le\delta\le\tilde{\delta}(\varepsilon)} a_k^\delta, & k < \tilde{k}(\varepsilon), \\ a^0 + \varepsilon, & k \ge \tilde{k}(\varepsilon), \end{cases}$$

and the sequence $\{\tilde{\gamma}_k^\varepsilon\}_{k\in\mathbb{N}_0}$ recursively via

$$\tilde{\gamma}_{k+1}^\varepsilon := \tilde{a}_k^\varepsilon + b\tilde{\gamma}_k^\varepsilon + c(\tilde{\gamma}_k^\varepsilon)^2, \qquad k \in \mathbb{N}_0, \qquad \tilde{\gamma}_0^\varepsilon := \gamma_0.$$

Then it follows by induction that

$$\gamma_k^\delta \le \tilde{\gamma}_k^\varepsilon, \qquad 0 \le k \le k_*(\delta),$$

provided that $0 \le \delta \le \tilde{\delta}(\varepsilon)$ which we will assume in the following. Moreover, it follows as above that

$$\tilde{\gamma}_k^\varepsilon \le \max\{\gamma_0, \underline{\gamma}(a)\} \le \overline{\gamma}(a) < \overline{\gamma}(a^0 + \varepsilon), \qquad k \in \mathbb{N}_0.$$

Let us first assume that $\tilde{\gamma}_k^\varepsilon < \underline{\gamma}(a^0 + \varepsilon)$ for all $k \ge \tilde{k}(\varepsilon)$. Then it follows with (4.52) that $\tilde{\gamma}_k^\varepsilon$ is strictly monotonically increasing for $k \ge \tilde{k}(\varepsilon)$.

Let us now assume that $\underline{\gamma}(a^0 + \varepsilon) \le \tilde{\gamma}_k^\varepsilon \le \overline{\gamma}(a^0 + \varepsilon)$ for some $k = l \ge \tilde{k}(\varepsilon)$. Then it again follows with (4.52) that

$$\begin{aligned} \underline{\gamma}(a^0 + \varepsilon) &= (a^0 + \varepsilon) + b\underline{\gamma}(a^0 + \varepsilon) + c\underline{\gamma}(a^0 + \varepsilon)^2 \;\le\; \tilde{a}_k^\varepsilon + b\tilde{\gamma}_k^\varepsilon + c(\tilde{\gamma}_k^\varepsilon)^2 \\ &= \tilde{\gamma}_{k+1}^\varepsilon \;\le\; \tilde{\gamma}_k^\varepsilon \;\le\; \overline{\gamma}(a^0 + \varepsilon). \end{aligned}$$

Hence, $\underline{\gamma}(a^0 + \varepsilon) \le \tilde{\gamma}_k^\varepsilon \le \overline{\gamma}(a^0 + \varepsilon)$ for all $k \ge l$ and $\tilde{\gamma}_k^\varepsilon$ is monotonically decreasing for $k \ge l$.

These arguments show that $\tilde{\gamma}_k^\varepsilon$ is convergent. Since $\overline{\gamma}(a) < \overline{\gamma}(a^0 + \varepsilon)$, the only possible limit then is $\underline{\gamma}(a^0 + \varepsilon)$. Hence,

$$\limsup_{\delta \to 0} \gamma_{k_*(\delta)}^\delta \le \underline{\gamma}(a^0 + \varepsilon), \qquad \limsup_{k \to \infty} \gamma_k^0 \le \underline{\gamma}(a^0 + \varepsilon).$$

Since $\varepsilon > 0$ was arbitrary, this proves the assertion.

Let us now assume that $c = 0$ and $b < 1$, i.e., we consider case (ii). It follows by induction that

$$\gamma_k^\delta \le b^k \gamma_0 + \sum_{j=0}^{k-1} a_j^\delta b^{k-1-j}, \qquad 0 \le k \le k_*(\delta).$$

The assertions now follow with the limsup condition satisfied by a_k^δ. □

It is well known that in the noisy case convergence can be only expected if the iteration is stopped appropriately. Besides the discrepancy principle (2.2), which is an a-posteriori stopping rule that was successfully used together with Landweber iteration (cf. Chapters 2 and 3), we first want to study an a-priori stopping rule. The reason is that even in the linear case $O(\sqrt{\delta})$ is the best possible rate that can be achieved for (4.36) with (2.2) (cf., e.g., [45]). However, with an a-priori rule the best rate will be $O(\delta^{\frac{2}{3}})$.

We will consider now the following a-priori stopping rule, where the iteration is stopped after $k_* = k_*(\delta)$ steps with

$$\begin{cases} \eta \alpha_{k_*}^{\mu+\frac{1}{2}} \le \delta < \eta \alpha_k^{\mu+\frac{1}{2}}, \qquad 0 \le k < k_*, \qquad 0 < \mu \le 1, \\ k_*(\delta) \to \infty \quad \text{and} \quad \eta \ge \delta \alpha_{k_*}^{-\frac{1}{2}} \to 0 \quad \text{as} \quad \delta \to 0, \quad \mu = 0, \end{cases} \tag{4.53}$$

for some $\eta > 0$. Note that, due to (4.41), this guarantees that $k_*(\delta) < \infty$, if $\delta > 0$ and that $k_*(\delta) \to \infty$ as $\delta \to 0$. In the noise free case ($\delta = 0$) we can set $k_*(0) := \infty$ and $\eta := 0$.

We will show in the next theorem that this a-priori stopping rule yields convergence and convergence rates for the iteratively regularized Gauss–Newton method (4.36) provided that $\|v\|$, c_R, c_Q, and η are sufficiently small.

Theorem 4.12 *Let Assumption 4.8 hold and let $k_* = k_*(\delta)$ be chosen according to* (4.53).

(i) If $0 \le \mu < 1/2$, we assume that the following closeness conditions hold:

$$\begin{cases} b + 2\sqrt{ac} \le 1 \quad \text{and} \quad \gamma_0 \le \overline{\gamma}(a), & c_Q > 0, \\ b < 1, & c_Q = 0, \end{cases}$$

$$\alpha_0^\mu((1+c_R+\tfrac{3}{2}c_Q c_\gamma \alpha_0^\mu)\|v\| + c_\gamma(c_R+\tfrac{3}{4}\rho c_Q)+\tfrac{1}{2}\eta) \le \rho\,,$$

where

$$\begin{aligned}
a &:= r^{\mu+\frac{1}{2}}(\|v\| + (1+c_R)(2c_R\|v\|+\eta))\,,\\
b &:= r^{\mu+\frac{1}{2}}(\tfrac{1}{2}c_Q\alpha_0^\mu((3+5c_R)\|v\|+\eta) + (1+c_R)(2c_R+\tfrac{3}{2}\rho c_Q))\,,\\
c &:= r^{\mu+\frac{1}{2}}c_Q\alpha_0^\mu(\tfrac{3}{2}c_Q\alpha_0^\mu\|v\| + c_R + \tfrac{3}{4}\rho c_Q)\,,\\
\gamma_0 &:= \alpha_0^{-(\mu+\frac{1}{2})}\|F'(x^\dagger)(x_0-x^\dagger)\|\,,\\
c_\gamma &:= \begin{cases} \max\{\gamma_0, \underline{\gamma}(a)\}\,, & c_Q>0\,,\\ \gamma_0 + \frac{a}{1-b}\,, & c_Q = 0\,,\end{cases}
\end{aligned}$$

and $\underline{\gamma}(a)$ and $\overline{\gamma}(a)$ are as in (4.51).

(ii) If $1/2 \le \mu \le 1$, we assume that the following closeness conditions hold:

$$b+2\sqrt{ac}\le 1\,, \qquad \gamma_0 \le \overline{\gamma}(a)\,, \quad \text{and} \quad \alpha_0^\mu c_\gamma \le \rho\,,$$

where

$$\begin{aligned}
a &:= r^\mu(\|v\| + \tfrac{1}{2}\eta)\,,\\
b &:= r^\mu L(\tfrac{1}{2}\alpha_0^{\mu-\frac{1}{2}}\|v\| + \|(F'(x^\dagger)^*F'(x^\dagger))^{\mu-\frac{1}{2}}v\|)\,,\\
c &:= r^\mu \tfrac{1}{4}L\alpha_0^{\mu-\frac{1}{2}}\,,\\
\gamma_0 &:= \alpha_0^{-\mu}\|x_0-x^\dagger\|\,,\\
c_\gamma &:= \max\{\gamma_0,\underline{\gamma}(a)\}\,,
\end{aligned}$$

and $\underline{\gamma}(a)$ and $\overline{\gamma}(a)$ are as in (4.51).

Then we obtain that

$$\|x_{k_*}^\delta - x^\dagger\| = \begin{cases} o(1)\,, & \mu = 0\,,\\ O\Big(\delta^{\frac{2\mu}{2\mu+1}}\Big), & 0<\mu\le 1\,.\end{cases}$$

For the noise free case ($\delta = 0$, $\eta = 0$) we obtain that

$$\|x_k - x^\dagger\| = \begin{cases} o(\alpha_k^\mu)\,, & 0\le\mu<1\,,\\ O(\alpha_k)\,, & \mu = 1\,,\end{cases}$$

and that

$$\|F(x_k)-y\| = \begin{cases} o\Big(\alpha_k^{\mu+\frac{1}{2}}\Big), & 0\le\mu<\frac{1}{2}\,,\\ O(\alpha_k)\,, & \frac{1}{2}\le\mu\le 1\,.\end{cases}$$

Proof. Let K and e_k^δ be defined as in Lemma 4.10. Let us first assume that $0 \le \mu < 1/2$ and that the conditions in (i) hold. Moreover, let $\gamma_k^\delta := \alpha_k^{-(\mu+\frac{1}{2})} \|Ke_k^\delta\|$. We will show by induction that γ_k^δ satisfies an estimate (4.50) with b, c as in (i) and

$$a_k^\delta := r^{\mu+\frac{1}{2}}(1+2c_R(1+c_R))w_k(\mu+\tfrac{1}{2}) + (1+c_R)\delta\alpha_{k_*-1}^{-(\mu+\frac{1}{2})}$$

and that $x_k^\delta \in \mathcal{B}_\rho(x^\dagger)$ for all $0 \le k \le k_*$. Note that, due to (4.17), (4.18), and (4.53), a_k^δ satisfies the conditions in Lemma 4.11 with a as in (i) and

$$a^0 := \begin{cases} r^{\mu+\frac{1}{2}}(1+c_R)\eta\,, & \mu > 0\,, \\ 0\,, & \mu = 0\,, \end{cases}$$

and that, due to Assumption 4.8 (i), $x_0 \in \mathcal{B}_\rho(x^\dagger)$. Let us assume that $x_j^\delta \in \mathcal{B}_\rho(x^\dagger)$ for all $0 \le j \le k$ and that

$$\gamma_{j+1}^\delta \le a_j^\delta + b\gamma_j^\delta + c(\gamma_j^\delta)^2 \tag{4.54}$$

for all $0 \le j < k$ and some $0 \le k < k_*$. Then it follows with (4.17), (4.41), and (4.45) that (4.54) is also satisfied for $j = k$. Due to the first two closeness conditions in (i), it follows as in Lemma 4.11 that

$$\gamma_k^\delta \le c_\gamma\,.$$

Together with (4.17), (4.41), (4.44), and (4.53) we obtain the estimate

$$\|e_{k+1}^\delta\| \le \alpha_0^\mu((1+c_R+\tfrac{3}{2}c_Qc_\gamma\alpha_0^\mu)\|v\|c_\gamma(c_R+\tfrac{3}{4}\rho c_Q)+\tfrac{1}{2}\eta)\,.$$

Due to the third closeness condition in (i) this implies that $x_{k+1}^\delta \in \mathcal{B}_\rho(x^\dagger)$. Due to (4.53), these induction steps are possible up to $k = k_* - 1$.

Now (4.18), (4.41), (4.44), Lemma 4.11 and (4.53) imply that

$$\|e_{k_*}^\delta\| = \begin{cases} O(\alpha_{k_*}^\mu) = O\Big(\big(\alpha_{k_*}^{\mu+\frac{1}{2}}\big)^{\frac{2\mu}{2\mu+1}}\Big) = O\Big(\delta^{\frac{2\mu}{2\mu+1}}\Big), & \mu > 0\,, \\ o(1)\,, & \mu = 0\,. \end{cases}$$

Note that $a^0 = 0$ if $\mu = 0$ and that in all interesting cases $0 < a$. The case $a = a^0 = 0$ corresponds to the trivial noise free case with $x_k = x^\dagger$.

For the general noise free case, $\delta = \eta = 0$, the assertions follow as above with Lemma 4.11 noting that then $a^0 = 0$ also for $\mu > 0$. The last assertion on $\|F(x_k) - y\|$ follows together with (4.42).

Let us now assume that $1/2 \le \mu \le 1$ and that the conditions in (ii) hold. Moreover, let $\gamma_k^\delta := \alpha_k^{-\mu}\|e_k^\delta\|$. Using (4.17), (4.41), (4.46), and (4.53), it follows with induction as in case (i) that γ_k^δ satisfies an estimate (4.50) with b, c as in (ii) and

$$a_k^\delta := r^\mu (w_k(\mu) + \tfrac{1}{2}\eta)$$

and that $x_k^\delta \in \mathcal{B}_\rho(x^\dagger)$ for all $0 \le k \le k_*$. Note that, due to (4.18), a_k^δ satisfies the conditions in Lemma 4.11 with a as in (ii) and

$$a^0 := \begin{cases} \frac{1}{2} r^\mu \eta\,, & \frac{1}{2} \le \mu < 1\,, \\ a\,, & \mu = 1\,. \end{cases}$$

Due to Lemma 4.11 we now obtain as in case (i) that

$$\|e_{k_*}^\delta\| = O\Big(\delta^{\frac{2\mu}{2\mu+1}}\Big) \quad \text{and} \quad \|e_k\| = \begin{cases} o(\alpha_k^\mu)\,, & \frac{1}{2} \le \mu < 1\,, \\ O(\alpha_k)\,, & \mu = 1\,. \end{cases}$$

Note that for the non-trivial noise free case $0 = a^0 < a$ if $\mu < 1$. Together with (4.47) we now obtain that $\|Ke_k\| = O(\alpha_k)$ and, hence together with the estimate $\|F(x_k) - y\| \le \|Ke_k^\delta\| + L\|e_k^\delta\|^2/2$ that $\|F(x_k) - y\| = O(\alpha_k)$. □

The closeness conditions in the theorem above cannot be checked for a real practical problem, since they involve the exact solution. One can see, however, that the conditions are satisfied if $\|v\|$, c_R, c_Q, and η are sufficiently small.

Since α_k can be chosen as $\alpha_k := 2^{-k}$, the iteratively regularized Gauss–Newton method converges much faster than the Landweber iteration method. However, each iteration step is more expensive due to the fact that one has to invert the operator $(F'(x_k^\delta)^* F'(x_k^\delta) + \alpha_k I)$.

Unfortunately, the a-priori stopping rule (4.53) has the disadvantage that one needs information on the smoothness of the exact solution, i.e., the parameter μ. A-posteriori stopping rules only need available information. Thus, we will now study the discrepancy principle (2.2), i.e., $k_* = k_*(\delta)$ is chosen such that

$$\|y^\delta - F(x_{k_*}^\delta)\| \le \tau\delta < \|y^\delta - F(x_k^\delta)\|\,, \qquad 0 \le k < k_*\,,$$

with $\tau > 1$ sufficiently large. As mentioned above already, one cannot expect to obtain a better rate than $O(\sqrt{\delta})$ with this selection criterion even if (2.22) is satisfied with $\mu > 1/2$.

As for the a-priori stopping rule, (4.53), we can prove convergence rates for $\mu \le 1/2$ if $\|v\|$, c_R, and c_Q are sufficiently small.

Theorem 4.13 *Let Assumption 4.8 hold for some $0 \le \mu \le 1/2$, where we assume that* (4.37) – (4.39) *also hold for $\mu = 1/2$, and let $k_* = k_*(\delta)$ be chosen according* *to (2.2) with $\tau > 1$. Moreover, we assume that the following closeness conditions hold:*

$$\begin{cases} b + 2\sqrt{ac} \le 1 \quad \textit{and} \quad \gamma_0 \le \overline{\gamma}(a)\,, & c_Q > 0\,, \\ b < 1\,, & c_Q = 0\,, \end{cases}$$

$$\alpha_0^{\mu}((1 + c_R + \tfrac{3}{2} c_Q c_\gamma \alpha_0^{\mu})\,\|v\| + c_\gamma(c_R + \tfrac{3}{4}\rho c_Q + \beta)) \le \rho\,,$$

$$rq < 1\,,$$

$$\frac{1 + 2c_R(1 + c_R)}{1 - rq} + (1 + \alpha_0^{-1}\|F'(x^\dagger)\|^2) r q^2 < 2\,,$$

where

$$\begin{aligned}
a &:= r^{\mu+\frac{1}{2}}(1 + 2c_R(1 + c_R))\,\|v\|\,,\\
b &:= r^{\mu+\frac{1}{2}}(\tfrac{1}{2} c_Q \alpha_0^{\mu}(3 + 5c_R)\,\|v\| + (1 + c_R)(2c_R + \tfrac{3}{2}\rho c_Q + 2\beta))\,,\\
c &:= r^{\mu+\frac{1}{2}} c_Q \alpha_0^{\mu}(\tfrac{3}{2} c_Q \alpha_0^{\mu}\,\|v\| + c_R + \tfrac{3}{4}\rho c_Q + \beta)\,,\\
\beta &:= \frac{1 + c_R + \frac{1}{2}\rho c_Q}{2(\tau - 1)}\,,\\
\gamma_0 &:= \alpha_0^{-(\mu+\frac{1}{2})}\,\|F'(x^\dagger)(x_0 - x^\dagger)\|\,,\\
c_\gamma &:= \begin{cases} \max\{\gamma_0, \underline{\gamma}(a)\}\,, & c_Q > 0\,,\\ \gamma_0 + \frac{a}{1-b}\,, & c_Q = 0\,, \end{cases}\\
q &:= r^{-(\mu+\frac{1}{2})} b + c_Q c_\gamma \alpha_0^{\mu}(\tfrac{3}{2} c_Q \alpha_0^{\mu}\,\|v\| + c_R + \tfrac{3}{4}\rho c_Q + \beta)\,,
\end{aligned}$$

and $\underline{\gamma}(a)$ and $\overline{\gamma}(a)$ are as in (4.51). Then we obtain the rates

$$\|x_{k_*}^{\delta} - x^\dagger\| = \begin{cases} o\left(\delta^{\frac{2\mu}{2\mu+1}}\right), & 0 \le \mu < \frac{1}{2}\,,\\ O(\sqrt{\delta})\,, & \mu = \frac{1}{2}\,. \end{cases}$$

Proof. Let K and e_k^{δ} be defined as in Lemma 4.10. We may assume that $v \ne 0$ in Assumption 4.8 (ii), because otherwise (1.2), (2.2), and $\tau > 1$ would imply that $x_{k_*}^{\delta} = x_0 = x^\dagger$.

Let $\gamma_k^{\delta} := \alpha_k^{-(\mu+\frac{1}{2})}\,\|K e_k^{\delta}\|$. Then one can show similar to the proof of Theorem 4.12 (case (i)) that γ_k^{δ} satisfies an estimate (4.50) with b, c as above and

$$a_k^{\delta} := r^{\mu+\frac{1}{2}}(1 + 2c_R(1 + c_R)) w_k(\mu + \tfrac{1}{2})$$

and that $x_k^\delta \in \mathcal{B}_\rho(x^\dagger)$ for all $0 \le k \le k_*$ noting that, due to (1.2), (2.2), and (4.42), the estimate

$$(\tau - 1)\delta < (1 + c_R + \tfrac{1}{2}\rho c_Q)\, \|Ke_k^\delta\|, \quad 0 \le k < k_*\,, \tag{4.55}$$

holds and that, due to (4.17) and (4.18), a_k^δ satisfies the conditions in Lemma 4.11 with a as above and

$$a^0 := \begin{cases} 0, & \mu < \frac{1}{2}, \\ a, & \mu = \frac{1}{2}. \end{cases}$$

Thus,

$$\gamma_k^\delta \le c_\gamma\,, \qquad 0 \le k \le k_*\,. \tag{4.56}$$

Now (4.41), (4.17), (4.44), and (4.55) imply that

$$\|e_{k_*}^\delta\| = O\Big(\alpha_{k_*-1}^\mu (w_{k_*-1}(\mu) + w_{k_*-1}(\mu + \tfrac{1}{2}) + \gamma_{k_*-1}^\delta)\Big). \tag{4.57}$$

We will show that all three terms in the estimate above converge with the desired rate. In a first step we show that

$$\|Ke_{k+1}^\delta\| \le \Big(\frac{1 + 2c_R(1 + c_R)}{1 - rq} + (1 + \alpha_0^{-1}\|K\|^2)q\Big)\alpha_k^{\mu+\frac{1}{2}} w_k(\mu + \tfrac{1}{2}) \tag{4.58}$$

for all $0 \le k < k_*$. Note that (4.41), (4.17), (4.45) and (4.56) yield the estimate

$$\begin{aligned} \|Ke_{k+1}^\delta\| \;\le\;& (1 + 2c_R(1 + c_R))\alpha_k^{\mu+\frac{1}{2}} w_k(\mu + \tfrac{1}{2}) + (1 + c_R + \tfrac{1}{2}c_Q c_\gamma \alpha_0^\mu)\delta \\ &+ \|Ke_k^\delta\| \Big(\tfrac{3}{2}(1 + c_R)c_Q\alpha_0^\mu \|v\| + c_Q c_R \alpha_0^\mu \|v\| + \tfrac{3}{2}c_Q^2 c_\gamma \alpha_0^{2\mu} \|v\| \\ &+ (2c_R + \tfrac{3}{2}\rho c_Q)(1 + c_R + \tfrac{1}{2}c_Q c_\gamma \alpha_0^\mu)\Big). \end{aligned}$$

Together with (4.55) we obtain that

$$\|Ke_{k+1}^\delta\| \le (1 + 2c_R(1 + c_R))\alpha_k^{\mu+\frac{1}{2}} w_k(\mu + \tfrac{1}{2}) + q\,\|Ke_k^\delta\| \tag{4.59}$$

and hence by induction that

$$\|Ke_{k+1}^\delta\| \le (1+2c_R(1+c_R)) \sum_{j=0}^{k} \alpha_j^{\mu+\frac{1}{2}} w_j(\mu+\tfrac{1}{2})q^{k-j} + \|Ke_0\|\,q^{k+1}. \tag{4.60}$$

Let $\{E_\lambda\}$ denote a spectral family of K^*K. (4.41) and (4.17) imply that

$$\begin{aligned} \alpha_{k+1}^{\mu+\frac{1}{2}} w_{k+1}(\mu + \tfrac{1}{2}) \;&=\; \Big(\int_0^\infty \Big(\frac{\alpha_{k+1}}{\alpha_{k+1} + \lambda}\Big)^2 \lambda^{2\mu+1}\, d\|E_\lambda v\|^2\Big)^{\frac{1}{2}} \\ &\ge\; r^{-1}\alpha_k^{\mu+\frac{1}{2}} w_k(\mu + \tfrac{1}{2}) \end{aligned}$$

and hence that

$$\alpha_j^{\mu+\frac{1}{2}} w_j(\mu+\tfrac{1}{2}) \le r^{k-j} \alpha_k^{\mu+\frac{1}{2}} w_k(\mu+\tfrac{1}{2}), \qquad 0 \le j \le k. \tag{4.61}$$

This together with (4.60) and

$$\begin{aligned} \|Ke_0\| &= \Big(\int_0^\infty \Big(\frac{\alpha_k+\lambda}{\alpha_k}\Big)^2 \Big(\frac{\alpha_k}{\alpha_k+\lambda}\Big)^2 \lambda^{2\mu+1}\, d\|E_\lambda v\|^2 \Big)^{\frac{1}{2}} \\ &\le (1+\alpha_k^{-1}\|K\|^2)\alpha_k^{\mu+\frac{1}{2}} w_k(\mu+\tfrac{1}{2}) \\ &\le (1+\alpha_0^{-1}\|K\|^2 r^k)\alpha_k^{\mu+\frac{1}{2}} w_k(\mu+\tfrac{1}{2}) \end{aligned}$$

yields

$$\begin{aligned} \|Ke_{k+1}^\delta\| \le \Big(&(1+2c_R(1+c_R)) \sum_{j=0}^{k} (rq)^{k-j} \\ &+ (1+\alpha_0^{-1}\|K\|^2 r^k) q^{k+1} \Big) \alpha_k^{\mu+\frac{1}{2}} w_k(\mu+\tfrac{1}{2}). \end{aligned}$$

Since $r > 1$ and $rq < 1$, this proves (4.58).

An inspection of the proof of Lemma 4.10 shows that analogously to the estimate (4.59) one obtains that

$$(1-2c_R(1+c_R))\alpha_k^{\mu+\frac{1}{2}} w_k(\mu+\tfrac{1}{2}) \le \|Ke_{k+1}^\delta\| + q\|Ke_k^\delta\|$$

for $0 \le k < k_*$. This together with (4.58) and (4.61) implies that

$$\begin{aligned} \|Ke_{k_*}^\delta\| &\ge (1-2c_R(1+c_R))\alpha_{k_*-1}^{\mu+\frac{1}{2}} w_{k_*-1}(\mu+\tfrac{1}{2}) \\ &\quad - q\Big(\frac{1+2c_R(1+c_R)}{1-rq} + (1+\alpha_0^{-1}\|K\|^2)q \Big) \alpha_{k_*-2}^{\mu+\frac{1}{2}} w_{k_*-2}(\mu+\tfrac{1}{2}) \\ &\ge \Big(2 - \frac{1+2c_R(1+c_R)}{1-rq} \\ &\quad - (1+\alpha_0^{-1}\|F'(x^\dagger)\|^2) rq^2 \Big) \alpha_{k_*-1}^{\mu+\frac{1}{2}} w_{k_*-1}(\mu+\tfrac{1}{2}). \end{aligned}$$

Since $b < 1$, we also have that $c_R + \rho c_Q/2 < 1$. Therefore, using (1.2), (2.2), (4.43), $\|e_{k_*}^\delta\| \le \rho$ and the fifth closeness condition, we get

$$\alpha_{k_*-1}^{\mu+\frac{1}{2}} w_{k_*-1}(\mu+\tfrac{1}{2}) = O(\|Ke_{k_*}^\delta\|) = O(\delta). \tag{4.62}$$

Due to (4.41) and (4.17), this implies that

$$k_* = k_*(\delta) \to \infty \quad \text{as} \quad \delta \to 0\,. \tag{4.63}$$

This together with (4.41), (4.17), (4.62), and the Hölder inequality yields

$$\begin{aligned}
&\alpha_{k_*-1}^{\mu} w_{k_*-1}(\mu) \\
&\quad = \Big(\int_0^\infty \Big(\Big(\frac{\alpha_{k_*-1}}{\alpha_{k_*-1}+\lambda}\Big)^2 \lambda^{2\mu+1}\Big)^{\frac{2\mu}{2\mu+1}} \Big(\frac{\alpha_{k_*-1}}{\alpha_{k_*-1}+\lambda}\Big)^{\frac{1}{2\mu+1}} \, d\|E_\lambda v\|^2\Big)^{\frac{1}{2}} \\
&\quad \le (\alpha_{k_*-1}^{\mu+\frac{1}{2}} w_{k_*-1}(\mu+\tfrac{1}{2}))^{\frac{2\mu}{2\mu+1}} \Big(\int_0^\infty \Big(\frac{\alpha_{k_*-1}}{\alpha_{k_*-1}+\lambda}\Big)^2 \, d\|E_\lambda v\|^2\Big)^{\frac{1}{4\mu+2}} \\
&\quad = o(\delta^{\frac{2\mu}{2\mu+1}})\,.
\end{aligned}$$

Moreover, (4.58) and (4.62) imply that

$$\|K e_{k_*-1}^\delta\| = O(\delta)\,.$$

Together with (4.57) and (4.62) we finally obtain that

$$\begin{aligned}
\|e_{k_*}^\delta\| &= o\Big(\delta^{\frac{2\mu}{2\mu+1}}\Big) + O\Big(\Big(\alpha_{k_*-1}^{\mu+\frac{1}{2}} w_{k_*-1}(\mu+\tfrac{1}{2})\Big)^{\frac{2\mu}{2\mu+1}} w_{k_*-1}(\mu+\tfrac{1}{2})^{\frac{1}{2\mu+1}} \\
&\qquad + \|K e_{k_*-1}^\delta\|^{\frac{2\mu}{2\mu+1}} (\gamma_{k_*-1}^\delta)^{\frac{1}{2\mu+1}}\Big) \\
&= o\Big(\delta^{\frac{2\mu}{2\mu+1}}\Big) + O\Big(\delta^{\frac{2\mu}{2\mu+1}} (w_{k_*-1}(\mu+\tfrac{1}{2}) + \gamma_{k_*-1}^\delta)^{\frac{1}{2\mu+1}}\Big)
\end{aligned}$$

The assertions now follow together with (4.17), (4.18), (4.56), and (4.63) noting that, due to Lemma 4.11,

$$\gamma_{k_*}^\delta = o(1)\,, \qquad 0 \le \mu < \tfrac{1}{2}\,,$$

since $a^0 = 0 < a$ for $\mu < 1/2$. □

4.3 Generalizations of the iteratively regularized Gauss–Newton method

In this section, we will study generalizations of the results in the previous sections into several directions:

First of all, we will consider regularization methods other than Tikhonov regularization for the linear subproblems in each Newton step. Such modified Newton type methods for solving (1.1) have been studied and analyzed in several recent publications, see, e.g., [37, 64, 79, 83, 135]. The in general unbounded inverse of $F'(x)$ in the definition of the Newton step is replaced by a bounded approximation, defined via a regularizing operator, i.e.,

$$R_\alpha(F'(x)) \approx F'(x)^\dagger .$$

More precisely, $\alpha > 0$ is a small regularization parameter, and R_α satisfies

$$\begin{aligned} &R_\alpha(K)y \to K^\dagger y \quad \text{as} \quad \alpha \to 0 \quad \text{for all} \quad y \in \mathcal{R}(K)\,, \\ &\|R_\alpha(K)\| \le \Phi(\alpha)\,, \quad \text{and} \quad \|R_\alpha(K)K\| \le c_K\,, \\ &\text{for all} \quad K \in \mathcal{L}(\mathcal{X}, \mathcal{Y}) \quad \text{with} \quad \|K\| \le c_s\,, \end{aligned} \tag{4.64}$$

for some positive function $\Phi(\alpha)$ and some positive constants c_K, c_s. Note that, especially in view of operators K with unbounded inverses, $\Phi(\alpha)$ has to tend to infinity as α goes to zero; we assume w.l.o.g. that $\Phi(\alpha)$ is proportional to $1/\sqrt{\alpha}$. A possibility of defining regularization methods satisfying (4.64) is given via spectral theory (cf., e.g., [45]) by choosing a piecewise continuous real function $g_\alpha : [0, \overline{\lambda}] \to \mathbb{R}$, $\overline{\lambda} > 0$, satisfying

$$\begin{aligned} &g_\alpha(\lambda) \to \lambda^{-1} \qquad \text{as} \quad \alpha \to 0 \quad \text{for all} \quad \lambda \in (0, \overline{\lambda}]\,, \\ &\sup_{\lambda \in [0,\overline{\lambda}]} |\lambda g_\alpha(\lambda)| \le c_g\,, \qquad \text{and} \qquad \sup_{\lambda \in [0,\overline{\lambda}]} |g_\alpha(\lambda)| \le c(\alpha)\,, \end{aligned} \tag{4.65}$$

for some positive constant c_g and some positive function $c(\alpha)$, and by setting

$$R_\alpha(K) := g_\alpha(K^*K)K^*. \tag{4.66}$$

Then R_α satisfies (4.64) with $\Phi(\alpha) = (c_g c(\alpha))^{\frac{1}{2}}$, $c_K = c_g$, and $c_s = \overline{\lambda}^{\frac{1}{2}}$.

Within the so-defined class, many well-known regularization methods such as Tikhonov regularization, Landweber iteration, and Lardy's method can be found. Note, however, that the slightly more general concept (4.64) additionally includes regularization by discretization, which we will also take into consideration below.

Using a monotonically decreasing sequence $\alpha_k \searrow 0$ of regularization parameters, one arrives at a class of regularized Newton methods

$$x_{k+1}^\delta = x_0 + R_{\alpha_k}(F'(x_k^\delta))(y^\delta - F(x_k^\delta) - F'(x_k^\delta)(x_0 - x_k^\delta))\,. \tag{4.67}$$

As in the previous sections, y^δ are noisy data satisfying the estimate (1.2) and the superscript δ is omitted in the noise free case. Note that in the limiting case $\alpha_k \to 0$, so that $R_{\alpha_k}(F'(x))$ converges pointwise towards $F'(x)^\dagger$, this formulation is equivalent to the usual Newton method. The special choice

$$R_{\alpha_k}(F'(x_k^\delta)) = (F'(x_k^\delta)^* F'(x_k^\delta) + \alpha_k I)^{-1} F'(x_k^\delta)^*$$

corresponds to the iteratively regularized Gauss–Newton method discussed in Section 4.2.

As a further generalization, we will not only consider Hölder type source conditions (2.22), but also logarithmic source conditions that are more appropriate for severely ill-posed problems (cf. [77, 79]), i.e.,

$$\begin{aligned} x^\dagger - x_0 &= f_\mu^L(F'(x^\dagger)^* F'(x^\dagger))v\,, \quad \mu > 0\,, \quad v \in \mathcal{N}(F'(x^\dagger))^\perp \subset \mathcal{X}\,, \\ f_\mu^L(\lambda) &:= (-\ln(\lambda c_L^{-1}))^{-\mu}\,, \qquad c_L > c_s^2\,, \end{aligned} \tag{4.68}$$

with c_s as in (4.64). As in the Hölder type case $f_\mu^L(K^*K)$ is defined via functional calculus (cf., e.g., [45]).

Note that a generalization to logarithmic source conditions is possible also for the nonlinear Landweber iteration discussed in Chapter 2, see [37].

In order to obtain convergence rates under regularity conditions (2.22) or (4.68), in analogy to the assertions of Lemma 4.5 for Tikhonov regularization with source conditions (2.22), the regularizing operators R_α have to converge to the inverse of K at some rate on the set of solutions satisfying the regularity conditions. The appropriate conditions are

$$\|(I - R_\alpha(K)K)(K^*K)^\mu\| \le c_{1,\mu}\alpha^\mu \quad \text{and} \tag{4.69}$$

$$\|K(I - R_\alpha(K)K)(K^*K)^\mu\| \le c_{2,\mu}\alpha^{\mu+\frac{1}{2}} \tag{4.70}$$

$$\text{for all} \quad K \in \mathcal{L}(\mathcal{X},\mathcal{Y}) \quad \text{with} \quad \|K\| \le c_s$$

in the Hölder type case and

$$\|(I - R_\alpha(K)K)f_\mu^L(K^*K)\| \le c_{3,\mu}|\ln(\alpha)|^{-\mu} \quad \text{and} \tag{4.71}$$

$$\|K(I - R_\alpha(K)K)f_\mu^L(K^*K)\| \le c_{4,\mu}\alpha^{\frac{1}{2}}|\ln(\alpha)|^{-\mu} \tag{4.72}$$

$$\text{for all} \quad K \in \mathcal{L}(\mathcal{X},\mathcal{Y}) \quad \text{with} \quad \|K\| \le c_s$$

in the logarithmic case. Here, $c_{1,\mu}$, $c_{2,\mu}$, $c_{3,\mu}$, $c_{4,\mu}$ are positive constants and $\alpha \le \overline{\alpha}$ for some $\overline{\alpha} < 1$ in (4.71).

To make all these methods well defined, we assume the forward operators F to be Fréchet-differentiable with derivatives being uniformly bounded in a neighbourhood of x_0. This uniform bound has to be such that applicability of the respective regularization method can be guaranteed. We assume that

$$\|F'(x)\| \le c_s \qquad \text{for all } x \in \mathcal{B}_{2\rho}(x_0) \subset \mathcal{D}(F)\,, \tag{4.73}$$

with c_s as in (4.64). This can always be achieved by proper scaling.

Newton type methods rely on the local validity of the first order Taylor approximation, i.e., the linearization of the nonlinear operator equation to be solved. In classical convergence proofs of Newton's method, a local Lipschitz condition on F',

$$\|F'(\tilde{x}) - F'(x)\| \leq L\|\tilde{x} - x\|, \qquad x, \tilde{x} \in \mathcal{B}_{2\rho}(x_0), \tag{4.74}$$

is used implying a Taylor remainder estimate being quadratic in terms of the difference between x and $\tilde{x}$. As we have seen in Section 4.2, unless a source condition (2.22) with $\mu \geq 1/2$ is satisfied, stronger assumptions on F are required in a convergence analysis.

The nonlinearity conditions considered so far, see (2.4), (2.25), (4.10), (4.12), and (4.37) – (4.39) enable an estimate of the Taylor remainder in terms of differences between function values of the forward operator. Similar to the analysis of the iteratively regularized Gauss–Newton method (compare (4.37) – (4.39)), we assume that

$$\begin{aligned} F'(\tilde{x}) = R(\tilde{x}, x)F'(x) + Q(\tilde{x}, x), \qquad & \|R(\tilde{x}, x)F'(x)\| \leq c_s, \\ \|I - R(\tilde{x}, x)\| \leq c_R\|\tilde{x} - x\|, \text{ and } & \|Q(\tilde{x}, x)\| \leq c_Q\|F'(x^\dagger)(\tilde{x} - x)\| \end{aligned} \tag{4.75}$$

hold for $x, \tilde{x} \in \mathcal{B}_{2\rho}(x_0)$, where c_R and c_Q are nonnegative constants with $c_R + c_Q > 0$. Note that for investigating the general case, we here use a slightly stronger estimate on $\|I - R(\tilde{x}, x)\|$ than in (4.38). In case $c_Q = 0$, this means local range invariance of the adjoint of $F'(x)$.

Alternatively, we will show that the convergence proofs for the iteratively regularized Gauss–Newton method, and more generally, Newton type methods induced by regularization methods according to (4.64) can be carried out under local range invariance of $F'(x)$, namely

$$F'(\tilde{x}) = F'(x)R(\tilde{x}, x) \qquad \text{and} \qquad \|I - R(\tilde{x}, x)\| \leq c_R\|\tilde{x} - x\| \tag{4.76}$$

for $x, \tilde{x} \in \mathcal{B}_{2\rho}(x_0)$ and some positive constant c_R. This does not extend to the part of the convergence analysis for the a-posteriori stopping rule according to the discrepancy principle: it is intuitively clear that when using this principle, which is based on information on function values of F only, also a Taylor estimate in terms of values of F, and hence an assumption of the type (4.75), will be required for analyzing a Newton type iteration.

Condition (4.76) is closely related to the so-called affine covariant Lipschitz condition in [38], which will be considered in more detail in Section 4.4 below.

To illustrate condition (4.76), we give the following simple example of parameter estimation from exterior measurements.

Example 4.14 Let $\Omega \subset \mathbb{R}^d$ with $d \in \{1, 2, 3\}$ be some bounded domain with smooth boundary and consider the problem of estimating the coefficient c in

$$\begin{aligned} -\Delta u + cu &= f, \qquad \text{in } \Omega, \\ u &= g, \qquad \text{on } \partial\Omega, \end{aligned} \tag{4.77}$$

from measurements of u outside $\Omega^* \subset \Omega$, where $f \in L^2(\Omega)$ and $g \in H^{\frac{3}{2}}(\Omega)$. In this example, the forward operator F is the composition of the parameter-to-solution mapping G for (4.77) with an observation operator T, i.e.:

$$F := T \circ G : \mathcal{D}(F) \to L^2(\Omega \setminus \Omega^*),$$

$$\begin{array}{rcl} G : \mathcal{D}(F) & \to & H^2(\Omega) \\ c & \mapsto & G(c) := u(c), \end{array} \qquad \begin{array}{rcl} T : H^2(\Omega) & \to & L^2(\Omega \setminus \Omega^*) \\ u & \mapsto & u\big|_{\Omega \setminus \Omega^*}. \end{array}$$

One can show (cf. [32]) that there exists $\gamma > 0$ such that G and hence F is well defined on

$$\mathcal{D}(F) = \{c \in L^2(\Omega) : \|c - \hat{c}\|_{L^2} < \gamma \text{ for some } \hat{c} \in L^2(\Omega) \text{ with } \hat{c} \geq 0 \text{ a.e.}\}.$$

It can be argued that the Fréchet-derivative of F is given by $F' = T \circ G'$, where $G'(c)h = -A(c)^{-1}[h\, u(c)]$, for $c \in \mathcal{D}(F)$ and $h \in \mathcal{X} = L^2(\Omega)$, and

$$\begin{array}{rcl} A(c) : H^2(\Omega) \cap H_0^1(\Omega) & \to & L^2(\Omega) \\ u & \mapsto & -\Delta u + cu. \end{array}$$

Consequently, if

$$|u(c_0)(x)| \geq \kappa > 0 \qquad \text{for all} \quad x \in \Omega$$

holds for some $c_0 \in \mathcal{D}(F)$ so that $|u(c)(x)| \geq \kappa/2 > 0$ is satisfied for all $c \in \mathcal{B}_{2\rho}(c_0)$, $x \in \Omega$, and some $\rho > 0$ sufficiently small, then (4.76) holds with

$$R(\tilde{c}, c)h = (u(c))^{-1} A(c) A(\tilde{c})^{-1} [h\, u(\tilde{c})].$$

Note that the same reasoning would go through when considering measurements of u on the boundary of Ω^* only instead of the full dimensional set $\Omega \setminus \Omega^*$, i.e., T would be replaced by some trace type operator, mapping the solution $u(c)$ to the measured boundary values. However, then, in general, c would not be uniquely identifiable from these measurements. For further parameter identification problems of similar type see, e.g., [22].

In order to be able to carry out the estimates in the proof of Theorem 4.16 below, we have to make some additional assumptions on the regularization parameters α_k and the regularization methods R_α in view of the respective nonlinearity conditions (4.74), (4.75), or (4.76). All conditions needed are summarized in the following assumption.

Assumption 4.15 Let ρ be a positive number such that $\mathcal{B}_{2\rho}(x_0) \subset \mathcal{D}(F)$.

(i) The equation $F(x) = y$ has an x_0-minimum-norm solution $x^\dagger$ in $\mathcal{B}_\rho(x_0)$ satisfying the condition $x^\dagger - x_0 \in \mathcal{N}(F'(x^\dagger))^\perp$. Moreover, F is Fréchet-differentiable and satisfies the scaling property (4.73).

(ii) Let R_α be a regularization method satisfying (4.64), where $\Phi(\alpha)$ satisfies

$$\Phi(\alpha) = c_\Phi \alpha^{-\frac{1}{2}} \tag{4.78}$$

for some positive constant c_Φ.

(iii) The parameters α_k in the iteration procedure (4.67) satisfy condition (4.41), i.e.,

$$\alpha_k > 0\,, \qquad 1 \le \frac{\alpha_k}{\alpha_{k+1}} \le r\,, \qquad \lim_{k\to\infty} \alpha_k = 0\,,$$

for some $r > 1$. In case (4.68) holds, we will assume that $\alpha_0 < 1$.

(iv) Let one of the following three conditions hold:

(a) The nonlinearity condition (4.76) holds together with

$$\|R_\alpha(KR)KR - R_\alpha(K)K\| \le c_1 \|I - R\| \tag{4.79}$$

for all operators $K \in \mathcal{L}(\mathcal{X}, \mathcal{Y})$ and $R \in \mathcal{L}(\mathcal{X}, \mathcal{X})$ with $\|K\| \le c_s$, $\|KR\| \le c_s$ and $\|I - R\| \le c_I < 1$.
If the source condition (2.22) or (4.68) holds for some $\mu > 0$, we also assume that R_α satisfies (4.69) or (4.71), respectively.

(b) The nonlinearity condition (4.75) holds and R_α satisfies (4.69) and (4.70) for all $0 \le \mu \le \mu_0$ for some $\mu_0 \ge 1/2$. (Note that μ_0 is called the qualification of the regularization method, cf. [45]). Moreover, we assume that

$$\|KR_\alpha(K)\| \le c_2 \tag{4.80}$$

and that either $c_Q = 0$ in (4.75) or

$$\|R_\alpha(\tilde{K})\tilde{K} - R_\alpha(K)K\| \le c_1 \|\tilde{K} - K\| \alpha^{-\frac{1}{2}}, \tag{4.81}$$

$$\|K(R_\alpha(\tilde{K})\tilde{K} - R_\alpha(K)K)\| \le c_1 \|\tilde{K} - K\| \tag{4.82}$$

hold for all $K, \tilde{K} \in \mathcal{L}(\mathcal{X}, \mathcal{Y})$ with $\|K\| \le c_s$, $\|\tilde{K}\| \le c_s$.

If the source condition (2.22) holds, we assume that $0 < \mu \leq 1/2$. In the logarithmic case, i.e., when (4.68) holds, we assume that R_α satisfies (4.71), (4.72), and the conditions

$$\|R_\alpha(RK)RK - R_\alpha(K)K\| \leq c_1 \|I - R\|, \qquad (4.83)$$

$$\|K(R_\alpha(RK)RK - R_\alpha(K)K)\| \leq c_1 \alpha^{\frac{1}{2}} \|I - R\| \qquad (4.84)$$

for all operators $K \in \mathcal{L}(\mathcal{X}, \mathcal{Y})$ and $R \in \mathcal{L}(\mathcal{Y}, \mathcal{Y})$ with $\|K\| \leq c_s$, $\|KR\| \leq c_s$ and $\|I - R\| \leq c_I < 1$.

(c) The Lipschitz condition (4.74) holds. The solution $x^\dagger$ and the regularization method R_α satisfy (2.22) and (4.69) for some $\mu \geq 1/2$, respectively. In addition, R_α fulfills the condition

$$\|(R_\alpha(\tilde{K})\tilde{K} - R_\alpha(K)K)(K^*K)^{\frac{1}{2}}\| \leq c_1 \|\tilde{K} - K\| \qquad (4.85)$$

for all $K, \tilde{K} \in \mathcal{L}(\mathcal{X}, \mathcal{Y})$ with $\|K\|, \|\tilde{K}\| \leq c_s$.

Here c_1, c_2, and c_I are some positive constants and c_s is as in (4.64).

Now we shall state and prove convergence as $k \to \infty$ in the noiseless case, as well as convergence in the situation of noisy data, using an appropriate a-priori stopping rule. In the general case, convergence can be achieved if the stopping index $k_* = k_*(\delta)$ is chosen such that (compare (4.53))

$$k_* \to \infty \qquad \text{and} \qquad \eta \geq \delta \alpha_{k_*}^{-\frac{1}{2}} \to 0 \ \text{ as } \ \delta \to 0 . \qquad (4.86)$$

If additional source conditions hold, an appropriate choice is

$$\eta \alpha_{k_*}^{\mu + \frac{1}{2}} \leq \delta < \eta \alpha_k^{\mu + \frac{1}{2}}, \qquad 0 \leq k < k_* , \qquad (4.87)$$

in the Hölder type case, i.e., when (2.22) holds, and

$$\eta \alpha_{k_*}^{\frac{1}{2}} |\ln(\alpha_{k_*})|^{-\mu} \leq \delta < \eta \alpha_k^{\frac{1}{2}} |\ln(\alpha_k)|^{-\mu}, \qquad 0 \leq k < k_* , \qquad (4.88)$$

in the logarithmic case, i.e., when (4.68) holds. In all cases η is some sufficiently small positive constant.

Theorem 4.16 *Let Assumption 4.15 hold and let x_k^δ be defined by the sequence* (4.67). *Moreover, let η and $\|x_0 - x^\dagger\|$ be sufficiently small. Then, in the noise free case ($\delta = 0, \eta = 0$), the sequence x_k converges to $x^\dagger$ as $k \to \infty$. In case of noisy data and with the choice* (4.86)*, $x_{k_*}^\delta$ converges to $x^\dagger$ as $\delta \to 0$.*

If in addition the source condition (2.22) *or* (4.68) *holds with* $\|v\|$ *sufficiently small and if* $k_* = k_*(\delta)$ *is chosen according to the stopping rule* (4.87) *and* (4.88)*, respectively, then we obtain the convergence rates*

$$\|x^\delta_{k_*} - x^\dagger\| = \begin{cases} O\Big(\delta^{\frac{2\mu}{2\mu+1}}\Big), & \textit{in the Hölder type case}\,, \\ O((1+|\ln(\delta)|^{-\mu})\,, & \textit{in the logarithmic case}\,. \end{cases}$$

In the noise free case, we obtain the rates

$$\|x_k - x^\dagger\| = \begin{cases} O(\alpha_k^\mu)\,, & \textit{in the Hölder type case}\,, \\ O(|\ln(\alpha_k)|^{-\mu})\,, & \textit{in the logarithmic case}\,. \end{cases}$$

Proof. To derive a recursive error estimate, we assume that the current iterate x^δ_k is in $\mathcal{B}_\rho(x^\dagger)$ and that $k < k_*$. Note that $k_* = \infty$ if $\delta = 0$. This guarantees that $x^\dagger, x^\delta_k \in \mathcal{B}_{2\rho}(x_0)$. Moreover, as in the previous section, we set $e^\delta_k := x^\delta_k - x^\dagger$, $K := F'(x^\dagger)$, and $K_k := F'(x^\delta_k)$ and assume that $c_R\|e^\delta_k\| \le c_I$, so that the conditions of Assumption 4.15 are applicable to K_k. Then e^δ_{k+1} may be rewritten in the form

$$\begin{aligned} e^\delta_{k+1} &= (I - R_{\alpha_k}(K)K)(x_0 - x^\dagger) \\ &\quad + (R_{\alpha_k}(K)K - R_{\alpha_k}(K_k)K_k)(x_0 - x^\dagger) \\ &\quad - R_{\alpha_k}(K_k)(F(x^\delta_k) - F(x^\dagger) - K_k e^\delta_k) - R_{\alpha_k}(K_k)(y - y^\delta) \end{aligned} \tag{4.89}$$

and hence, due to (1.2), (4.64), and (4.78)

$$\begin{aligned} \|e^\delta_{k+1}\| &\le w_k + \|(R_{\alpha_k}(K)K - R_{\alpha_k}(K_k)K_k)(x_0 - x^\dagger)\| \\ &\quad + \|R_{\alpha_k}(K_k)(F(x^\delta_k) - F(x^\dagger) - K_k e^\delta_k)\| + c_\Phi \delta \alpha_k^{-\frac{1}{2}}\,, \end{aligned} \tag{4.90}$$

where

$$w_k := \|(I - R_{\alpha_k}(K)K)(x_0 - x^\dagger)\| \to 0 \qquad \text{as} \quad k \to \infty\,, \tag{4.91}$$

since $x_0 - x^\dagger \in \mathcal{N}(F'(x^\dagger))^\perp$ (see Assumption 4.15 (i)). In the Hölder type or logarithmic case, i.e., when (2.22) and (4.69) or (4.68) and (4.71) hold, respectiveley, by (4.73) we even obtain that

$$w_k \le \begin{cases} c_{1,\mu}\alpha_k^\mu \|v\|\,, & \text{in the Hölder type case}\,, \\ c_{3,\mu}|\ln(\alpha_k)|^{-\mu}\|v\|\,, & \text{in the logarithmic case}\,. \end{cases} \tag{4.92}$$

Let us first consider case (iv)(a) of Assumption 4.15. Due to (4.76), the third term on the right hand side of (4.89) can be rewritten as

$$R_{\alpha_k}(K_k)K_k \int_0^1 (R(x^\dagger + te^\delta_k, x^\delta_k) - I)e^\delta_k\,dt\,,$$

which together with (4.64), (4.73), (4.76), (4.79), and (4.90) yields

$$\|e_{k+1}^\delta\| \le w_k + c_1 c_R \|x_0 - x^\dagger\| \, \|e_k^\delta\| + \tfrac{1}{2} c_K c_R \|e_k^\delta\|^2 + c_\Phi \delta \alpha_k^{-\frac{1}{2}} . \tag{4.93}$$

Due to (4.41), (4.64), (4.86), and (4.91), we can now apply Lemma 4.11 to the sequence $\gamma_k^\delta := \|e_k^\delta\|$ with

$$\begin{aligned} &a_k^\delta := w_k + \delta \alpha_k^{-\frac{1}{2}} \quad \text{for } 0 \le k \le k_* , && a := \eta + \|x_0 - x^\dagger\| (1 + c_K) , \\ &a^0 := 0 , \qquad b := c_1 c_R \|x_0 - x^\dagger\| , && c := \tfrac{1}{2} c_K c_R . \end{aligned}$$

If η and $\|x_0 - x^\dagger\|$ are sufficiently small, then a, b, and c satisfy all the conditions of the lemma. Moreover, $\max\{\gamma_0, \underline{\gamma}(a)\}$ can be made smaller than $\min\{\rho, c_I/c_R\}$, so that we can guarantee that x_k^δ remains in $\mathcal{B}_\rho(x^\dagger)$ and $c_R \|e_k^\delta\| \le c_I$ for all $k \le k_*$. Thus, x_k^δ converges to $x^\dagger$ as $k \to \infty$ in the noise free case, and as $\delta \to 0$ in the noisy case, respectively.

To prove convergence rates under source conditions, we consider the sequence $\gamma_k^\delta := \psi(\alpha_k)^{-1} \|e_k^\delta\|$, where

$$\psi(\alpha) := \begin{cases} \alpha^\mu , & \text{in the Hölder type case} , \\ |\ln(\alpha)|^{-\mu} , & \text{in the logarithmic case} . \end{cases} \tag{4.94}$$

Then (4.41), (4.92), and (4.93) imply that

$$\gamma_{k+1}^\delta \le \tilde{c} (c_\mu \|v\| + c_1 c_R \|x_0 - x^\dagger\| \gamma_k^\delta + \tfrac{1}{2} c_K c_R \psi(\alpha_k) (\gamma_k^\delta)^2 + c_\Phi \delta \alpha_k^{-\frac{1}{2}} \psi(\alpha_k)^{-1}) ,$$

where $\tilde{c} := r^\mu$, $c_\mu := c_{1,\mu}$ in the Hölder type case and $\tilde{c} := (1 + \ln(r) |\ln(\alpha_0)|^{-1})^\mu$, $c_\mu := c_{3,\mu}$ in the logarithmic case, respectively.

Hence, due to (4.87) and (4.88), Lemma 4.11 is again applicable with

$$\begin{aligned} &a_k^\delta := \tilde{c} (c_\mu \|v\| + c_\Phi \delta \alpha_k^{-\frac{1}{2}} \psi(\alpha_k)^{-1}) \quad \text{for} \quad 0 \le k < k_* , \\ &a^0 := a := \tilde{c} (c_\mu \|v\| + c_\Phi \eta) , \quad b := \tilde{c} c_1 c_R \|x_0 - x^\dagger\| , \quad c := \tfrac{1}{2} \tilde{c} c_K c_R \psi(\alpha_0) , \end{aligned}$$

provided $\|v\|$ and η are sufficiently small. Thus, γ_k^δ is uniformly bounded for $k \le k_*$, and hence

$$\|e_k^\delta\| \le \tilde{\gamma} \psi(\alpha_k) , \qquad 0 \le k \le k_* ,$$

for some positive constant $\tilde{\gamma}$. This together with (4.94) immediately yields the convergence rate result in the noise free case. In the noisy case, the error estimate in terms of δ follows together with (4.87) and (4.88), respectively.

We now consider case (iv)(b) of Assumption 4.15. Note that, due to (4.75), the first two terms on the right hand side of the error decomposition (4.89) may be rearranged to

$$\begin{aligned}(I - R_{\alpha_k}(K)K)(x_0 - x^\dagger) &+ (R_{\alpha_k}(K)K - R_{\alpha_k}(K_k)K_k)(x_0 - x^\dagger) \\ &= (I - R_{\alpha_k}(R_k^\delta K)R_k^\delta K)(x_0 - x^\dagger) \\ &\quad + (R_{\alpha_k}(R_k^\delta K)R_k^\delta K - R_{\alpha_k}(K_k)K_k)(x_0 - x^\dagger)\,,\end{aligned}$$

where $R_k^\delta := R(x_k^\delta, x^\dagger)$, and that

$$\begin{aligned}F(x_k^\delta) - F(x^\dagger) - K_k e_k^\delta &= \int_0^1 [\,(R(x^\dagger + te_k^\delta, x^\dagger) - R_k^\delta)Ke_k^\delta \\ &\qquad + (Q(x^\dagger + te_k^\delta, x^\dagger) - Q(x_k^\delta, x^\dagger))e_k^\delta\,]\,dt\,.\end{aligned}$$

Thus, we obtain together with (4.64), (4.73), (4.75), (4.78), $c_Q = 0$ or (4.81), and (4.90) that

$$\begin{aligned}\|e_{k+1}^\delta\| &\le \|(I - R_{\alpha_k}(R_k^\delta K)R_k^\delta K)(x_0 - x^\dagger)\| \\ &\quad + c_1 c_Q \|x_0 - x^\dagger\| \alpha_k^{-\frac{1}{2}} \|Ke_k^\delta\| \\ &\quad + \tfrac{3}{2}(c_R + c_Q)c_\Phi \alpha_k^{-\frac{1}{2}} \|e_k^\delta\|\,\|Ke_k^\delta\| + c_\Phi \delta \alpha_k^{-\frac{1}{2}}\,.\end{aligned} \tag{4.95}$$

Since (4.69) holds,

$$\|(I - R_{\alpha_k}(R_k^\delta K)R_k^\delta K)(x_0 - x^\dagger)\| \to 0 \qquad \text{as} \quad k \to \infty\,. \tag{4.96}$$

This can be seen as follows: since, due to (4.64) and (4.75), the operators $(I - R_{\alpha_k}(R_k^\delta K)R_k^\delta K) : \mathcal{N}(K)^\perp \to X$ are uniformly bounded, by the Banach–Steinhaus Theorem it is sufficient to show that (4.96) holds for all $x_0 - x^\dagger$ from a dense subset of $\mathcal{N}(K)^\perp$, e.g., the set $\{(K^*K)^\mu v : v \in X\}$ for some $0 < \mu \le 1/2$. But this is satisfied, since

$$\|(I - R_{\alpha_k}(R_k^\delta K)R_k^\delta K)(K^*K)^\mu\| \le c_{1,\mu}(1 - c_I)^{-2\mu}\alpha_k^\mu \tag{4.97}$$

due to (4.13), (4.69), and (4.75).

If (4.83) holds, we obtain together with (4.75) and (4.91) that

$$\|(I - R_{\alpha_k}(R_k^\delta K)R_k^\delta K)(x_0 - x^\dagger)\| \le w_k + c_1 c_R \|x_0 - x^\dagger\|\,\|e_k^\delta\|$$

which together with (4.95) yields the estimate

$$\begin{aligned}\|e_{k+1}^\delta\| &\le w_k + c_1\|x_0 - x^\dagger\|(c_R\|e_k^\delta\| + c_Q \alpha_k^{-\frac{1}{2}}\|Ke_k^\delta\|) \\ &\quad + \tfrac{3}{2}(c_R + c_Q)c_\Phi \alpha_k^{-\frac{1}{2}} \|e_k^\delta\|\,\|Ke_k^\delta\| + c_\Phi \delta \alpha_k^{-\frac{1}{2}}\,.\end{aligned} \tag{4.98}$$

Since $\alpha_k^{-\frac{1}{2}}$ shows up in the second and third term of the estimates (4.95) and (4.98), we can only get a proper bound by making direct use of the fact that $\|Ke_k^\delta\|$ can be expected to converge faster to zero than $\|e_k^\delta\|$. For this purpose, analogously to the proof of Lemma 4.10, we derive a separate estimate also for $\|Ke_k^\delta\|$ by applying K to both sides of (4.89). Thus, we obtain together with (1.2), (4.64), (4.73), (4.75), (4.78), (4.80), and $c_Q = 0$ or (4.82) that

$$\begin{aligned}\|Ke_{k+1}^\delta\| \;\leq\;& \|K(I - R_{\alpha_k}(R_k^\delta K)R_k^\delta K)(x_0 - x^\dagger)\| \\ &+ c_1 c_Q (1-c_I)^{-1}\|x_0 - x^\dagger\|\,\|Ke_k^\delta\| \\ &+ \Big((1 + c_R\|e_k^\delta\|)c_2 + c_Q c_\Phi \alpha_k^{-\frac{1}{2}}\|Ke_k^\delta\|\Big) \\ &\cdot\Big(\tfrac{3}{2}(c_R + c_Q)\|e_k^\delta\|\,\|Ke_k^\delta\| + \delta\Big),\end{aligned} \tag{4.99}$$

where we used the representation $K = R(x^\dagger, x_k^\delta)K_k + Q(x^\dagger, x_k^\delta)$ and the estimate $\|(R_k^\delta)^{-1}\| \leq (1 - c_I)^{-1}$.

Again, similarly to (4.96) and (4.97), we obtain together with (4.13), (4.70), (4.75), and the Banach–Steinhaus Theorem that

$$\alpha_k^{-\frac{1}{2}}\|K(I - R_{\alpha_k}(R_k^\delta K)R_k^\delta K)(x_0 - x^\dagger)\| \to 0 \quad \text{as} \quad k \to \infty \tag{4.100}$$

and that

$$\|K(I - R_{\alpha_k}(R_k^\delta K)R_k^\delta K)(K^*K)^\mu\| \leq c_{2,\mu}(1 - c_I)^{-(2\mu+1)}\alpha_k^{\mu+\frac{1}{2}} \tag{4.101}$$

for some $0 < \mu \leq 1/2$.

If (4.84) holds, we obtain together with (4.75) that

$$\begin{aligned}\|K(I - R_{\alpha_k}(R_k^\delta K)R_k^\delta K)(x_0 - x^\dagger)\| \;\leq\;& \|K(I - R_{\alpha_k}(K)K)(x_0 - x^\dagger)\| \\ &+ c_1 c_R\|x_0 - x^\dagger\|\,\alpha_k^{\frac{1}{2}}\|e_k^\delta\|\end{aligned}$$

which together with (4.99) yields the estimate

$$\begin{aligned}\|Ke_{k+1}^\delta\| \;\leq\;& \|K(I - R_{\alpha_k}(K)K)(x_0 - x^\dagger)\| \\ &+ c_1\|x_0 - x^\dagger\|\,(c_R\alpha_k^{\frac{1}{2}}\|e_k^\delta\| + c_Q(1-c_I)^{-1}\|Ke_k^\delta\|) \\ &+ \Big((1 + c_R\|e_k^\delta\|)c_2 + c_Q c_\Phi \alpha_k^{-\frac{1}{2}}\|Ke_k^\delta\|\Big) \\ &\cdot\Big(\tfrac{3}{2}(c_R + c_Q)\|e_k^\delta\|\,\|Ke_k^\delta\| + \delta\Big).\end{aligned} \tag{4.102}$$

Note that

$$\|K(I - R_{\alpha_k}(K)K)(x_0 - x^\dagger)\| \le c_{4,\mu}\alpha_k^{\frac{1}{2}}|\ln(\alpha_k)|^{-\mu}\|v\| \qquad (4.103)$$

if (4.68) and (4.72) hold.

Similarly as above, we now apply Lemma 4.11 to

$$\gamma_k^\delta := \max\{\|e_k^\delta\|, \alpha_k^{-\frac{1}{2}}\|Ke_k^\delta\|\}$$

to show that x_k^δ remains in $\mathcal{B}_\rho(x^\dagger)$ and $c_R\|e_k^\delta\| \le c_I$ for all $k \le k_*$ and that x_k^δ converges to $x^\dagger$ as $k \to \infty$ in the noise free case, and as $\delta \to 0$ in the noisy case, respectively, if η and $\|x_0 - x^\dagger\|$ are sufficiently small, by combining (4.41), (4.86), (4.95), (4.96), (4.99), and (4.100).

The convergence rates are obtained by applying Lemma 4.11 to

$$\gamma_k^\delta := \psi(\alpha_k)^{-1}\max\{\|e_k^\delta\|, \alpha_k^{-\frac{1}{2}}\|Ke_k^\delta\|\},$$

where $\psi(\alpha)$ is as in (4.94). Here we have to combine (4.41), (4.87), (4.95), (4.97), (4.99), and (4.101) in the Hölder type case, i.e., when (2.22) holds for some $0 < \mu \le 1/2$. In the logarithmic case, i.e., when (4.68) holds, we have to combine (4.41), (4.88), (4.92), (4.98), (4.102), and (4.103). In all cases $\|v\|$ and η have to be sufficiently small.

Finally, we treat case (iv)(c) of Assumption 4.15. Using (2.22) with $\mu \ge 1/2$, (4.64), (4.73), (4.74), (4.78), (4.85), (4.90), and (4.92), we obtain the estimate

$$\|e_{k+1}^\delta\| \le c_{1,\mu}\alpha_k^\mu\|v\| + c_1 L\|K\|^{2\mu-1}\|v\|\,\|e_k^\delta\| + \tfrac{1}{2}Lc_\Phi\alpha_k^{-\frac{1}{2}}\|e_k^\delta\|^2 + c_\Phi\delta\alpha_k^{-\frac{1}{2}}.$$

As above, an application of Lemma 4.11 to $\gamma_k^\delta := \alpha_k^{-\mu}\|e_k^\delta\|$ yields the desired rates provided that $\|v\|$ and η are sufficiently small. □

Now we will apply this theorem to some special regularization methods R_α that are defined via spectral theory as in (4.66) with a function g_α satisfying (4.65). We consider three methods: Tikhonov regularization, iterated Tikhonov regularization, and Landweber iteration.

Tikhonov regularization is defined via $g_\alpha(\lambda) := (\lambda + \alpha)^{-1}$ yielding

$$R_\alpha(K) = (K^*K + \alpha I)^{-1}K^*, \quad I - R_\alpha(K)K = \alpha(K^*K + \alpha I)^{-1}. \qquad (4.104)$$

As mentioned in the beginning of this section, this choice yields the iteratively regularized Gauss–Newton method that was extensively studied in Section 4.2. A generalization for this method consists in treating logarithmic source conditions and the nonlinearity condition (4.76). The results concerning logarithmic source conditions can be found in Hohage's thesis [79].

It is well known (cf., e.g., [45]) that R_α as in (4.104) satisfies (4.64) and (4.78) with $c_K = c_\Phi = 1$. c_s can be chosen arbitrary; according to (4.73) the best choice then is $c_s := \sup_{x\in\mathcal{B}_{2\rho}(x_0)} \|F'(x)\|$.

Iterated Tikhonov regularization is defined via

$$g_\alpha(\lambda) := \sum_{j=0}^{n} \beta_j^{-1} \prod_{l=j}^{n} \beta_l(\lambda+\beta_l)^{-1},$$

where $\{\beta_j\}$ is a bounded sequence in $\mathbb{R}^+$ such that also $\beta_{j+1}^{-1}\beta_j$ is bounded. This yields

$$\begin{aligned} R_\alpha(K) &= \sum_{j=0}^{n} \beta_j^{-1}\Big(\prod_{l=j}^{n} \beta_l(K^*K+\beta_l I)^{-1}\Big)K^*, \\ I - R_\alpha(K)K &= \prod_{j=0}^{n} \beta_j(K^*K+\beta_j I)^{-1}. \end{aligned} \tag{4.105}$$

The calculation of $w_n := R_\alpha(K)z$ is done iteratively via

$$w_n = (K^*K+\beta_n I)^{-1}(K^*z+\beta_n w_{n-1}), \qquad w_{-1} := 0.$$

The effective regularization parameter α in (4.105) is given by

$$\alpha = \alpha_k := \Big(\sum_{j=0}^{n_k} \beta_j^{-1}\Big)^{-1}.$$

Thus, the method is only defined for a sequence of regularization parameters. The number of inner iterations, n_k, has to be chosen such that (4.41) (cf. Assumption 4.15 (iii)) is satisfied. In view of Theorem 4.16, α_k should decay as fast as possible so that less Newton steps are needed for fixed δ.

We will restrict the choice of β_j to the special sequence

$$\beta_j := \beta q^j, \tag{4.106}$$

with $q \in (0,1]$ and some positive constant β. If $q = 1$, the choice is stationary and becomes Lardy's method, otherwise it is non-stationary, which is more attractive, since less iteration steps are needed. For the effective regularization parameter it then holds that

$$\alpha_k = \beta(n_k+1)^{-1} \ \text{if} \ q=1 \qquad \text{and} \qquad \alpha_k \sim q^{n_k} \ \text{if} \ q<1. \tag{4.107}$$

This means, for Lardy's method n_k should grow exponentially, while in the non-stationary case $n_k \sim k$.

Iterated Tikhonov regularization also satisfies (4.64) and (4.78) with arbitrary c_s (cf. [65]).

Landweber iteration is defined via

$$g_\alpha(\lambda) := \sum_{j=0}^{n-1} (1-\lambda)^j$$

yielding

$$R_\alpha(K) = \sum_{j=0}^{n-1} (I - K^*K)^j K^*, \quad I - R_\alpha(K)K = (I - K^*K)^n, \tag{4.108}$$

with the effective regularization parameter

$$\alpha = \alpha_k := (n_k + 1)^{-1},$$

where, as for Lardy's method, n_k should grow exponentially.

It follows with Lemma 2.10 that this regularization method satisfies (4.64) and (4.78) with $c_K = c_\Phi = 1$ if $c_s = 1$. Note that one could also choose a larger scaling parameter, namely some $c_s < \sqrt{2}$. Then the estimates in Lemma 2.10 remain valid for some larger constants. However, we will assume in the following that $c_s = 1$.

Corollary 4.17 *Let Assumption 4.15 (i) and (iii) hold and let x_k^δ be defined by the sequence* (4.67)*, where the regularization method R_α is defined by either* (4.104), (4.105) *with* (4.106), *or* (4.108). *Moreover, let one of the following three conditions hold:*

(i) The nonlinearity condition (4.76) *holds.*

(ii) The nonlinearity condition (4.75) *holds and we assume that $0 < \mu \leq 1/2$ if the source condition* (2.22) *holds.*

(iii) The Lipschitz condition (4.74) *holds and the solution $x^\dagger$ satisfies* (2.22) *for some $\mu \geq 1/2$.*

Then the convergence and convergence rates results of Theorem 4.16 *hold for the appropriate smallness and source conditions as well as the appropriate stopping rules, where in case of Tikhonov regularization, i.e.,* (4.104), *$\mu \leq 1$ has to hold in* (2.22).

Proof. Since we have mentioned above that the three regularization methods under consideration satisfy conditions (4.64) and (4.78) and hence fulfill Assumption 4.15 (ii), the proof follows immediately with Theorem 4.16 if we can show that R_α satisfies the appropriate conditions of Assumption 4.15 (iv).

It is well known that all three methods satisfy (4.69), where in case of Tikhonov regularization the restriction $\mu \leq 1$ holds (cf. (4.17), [65], and (2.33), respectively). Due to the fact that the following symmetry condition

$$(K^*K)^{\frac{1}{2}}(R_\alpha(K)K) = R_\alpha(K)K(K^*K)^{\frac{1}{2}}$$

holds for all regularization methods constructed via (4.66), this already implies (4.70) with $c_{2,\mu} = c_{1,\mu+\frac{1}{2}}$, where in case of Tikhonov regularization the restriction $\mu \leq 1/2$ holds. Moreover, it also implies (4.71) and (4.72) (see [80, Lemma 4]). Since

$$KR_\alpha(K) = R_\alpha(K^*)K^*$$

holds for all regularization methods constructed via (4.66), condition (4.80) follows from (4.64) with $c_2 = c_K$. Since also

$$(R_\alpha(K)K)^* = R_\alpha(K)K$$

holds for all regularization methods constructed via (4.66), condition (4.85) follows from (4.82). Thus, it remains to show that the conditions (4.79) and (4.81) – (4.84) hold for the specific regularization methods.

First we show them for Tikhonov regularization, i.e., (4.104). Using the decomposition

$$\begin{aligned} R_\alpha(\tilde{K})\tilde{K} - R_\alpha(K)K &= \alpha(K^*K + \alpha I)^{-1}\Big(K^*(\tilde{K} - K) \\ &\quad + (\tilde{K}^* - K^*)\tilde{K}\Big)(\tilde{K}^*\tilde{K} + \alpha I)^{-1}, \end{aligned} \tag{4.109}$$

it immediately follows with (4.17) that (4.81) and (4.82) hold.

Let us now assume that $R \in \mathcal{L}(\mathcal{X}, \mathcal{X})$ with $\|I - R\| \leq c_I < 1$. Then (4.17) and (4.109) imply that

$$\|R_\alpha(KR)KR - R_\alpha(K)K\| \leq \|I - R\|(1 + \|R^{-1}\|)$$

and hence (4.79) is shown.

Finally, we assume that $R \in \mathcal{L}(\mathcal{Y}, \mathcal{Y})$ with $\|I - R\| \leq c_I < 1$ to obtain together with (4.17) and (4.109) that

$$\begin{aligned} \|R_\alpha(RK)RK - R_\alpha(K)K\| &\leq \tfrac{1}{4}\|I - R\|(1 + \|R^{-1}\|), \\ \|K(R_\alpha(RK)RK - R_\alpha(K)K)\| &\leq \tfrac{1}{2}\|I - R\|(1 + \|R^{-1}\|)\,\alpha^{\frac{1}{2}} \end{aligned}$$

which yields (4.83) and (4.84).

For the iterative methods, iterated Tikhonov regularization and Landweber iteration, we make use of the formulae

$$\prod_{j=0}^{n} A_j - \prod_{j=0}^{n} B_j = \sum_{j=0}^{n} \Big(\prod_{l=0}^{j-1} A_l\Big)(A_j - B_j)\Big(\prod_{l=j+1}^{n} B_l\Big), \tag{4.110}$$

$$\begin{aligned} &\sum_{j=m}^{p} \Big(\prod_{l=0}^{j-1} A_l\Big)(I - A_j)C\Big(\prod_{l=j+1}^{n} B_l\Big) \\ &\quad = \sum_{j=m}^{p} \Big(\prod_{l=0}^{j-1} A_l\Big)C(I - B_j)\Big(\prod_{l=j+1}^{n} B_l\Big) \\ &\qquad + \Big(\prod_{l=0}^{m-1} A_l\Big)C\Big(\prod_{l=m}^{n} B_l\Big) - \Big(\prod_{l=0}^{p} A_l\Big)C\Big(\prod_{l=p+1}^{n} B_l\Big) \end{aligned} \tag{4.111}$$

that hold for linear operators A_j, B_j, C for $0 \le m \le p \le n$, with the usual conventation $\prod_{j=0}^{-1} A_j := I$ and $\prod_{j=n+1}^{n} B_j := I$. The second formula follows with a telescope sum argument and the first formula is actually a special case of the second one.

Using (4.110) with $A_j := (I - K^*K)$ and $B_j := (I - \tilde{K}^*\tilde{K})$ we obtain for Landweber iteration the representation

$$\begin{aligned} R_\alpha(\tilde{K})\tilde{K} - R_\alpha(K)K &= (I - K^*K)^n - (I - \tilde{K}^*\tilde{K})^n \\ &= \sum_{j=0}^{n-1} (I - K^*K)^j \Big(K^*(\tilde{K} - K) \\ &\qquad + (\tilde{K}^* - K^*)\tilde{K}\Big)(I - \tilde{K}^*\tilde{K})^{n-1-j}. \end{aligned} \tag{4.112}$$

This together with (2.34) yields

$$\begin{aligned} \|R_\alpha(\tilde{K})\tilde{K} - R_\alpha(K)K\| &\le \|\tilde{K} - K\| \sum_{j=0}^{n-1} \Big((j+1)^{-\frac{1}{2}} + (n-j)^{-\frac{1}{2}}\Big) \\ &\le 4n^{\frac{1}{2}} \|\tilde{K} - K\|, \end{aligned}$$

hence (4.81) holds. Moreover, we obtain

$$\begin{aligned}
&\|K(R_\alpha(\tilde{K})\tilde{K} - R_\alpha(K)K)\| \\
&\quad \leq \Big\| \sum_{j=0}^{n-1} (I - KK^*)^j KK^*(\tilde{K} - K)(I - \tilde{K}^*\tilde{K})^{n-1-j} \Big\| \\
&\quad\quad + \|\tilde{K} - K\| \sum_{j=0}^{n-1} (j+1)^{-\frac{1}{2}}(n-j)^{-\frac{1}{2}}.
\end{aligned}$$

Since, due to Lemma 2.9, the sum in the second term is bounded, (4.82) is shown to hold if we can show that also the first term can be estimated from above by $\|\tilde{K} - K\|$ times a constant. To achieve this we split the sum into two parts and use formula (4.111) with $A_j := (I - KK^*)$, $B_j := (I - \tilde{K}^*\tilde{K})$, $C := \tilde{K} - K$, and $m := 0$ as well as (2.33) to obtain

$$\begin{aligned}
&\Big\| \sum_{j=0}^{n-1} (I - KK^*)^j KK^*(\tilde{K} - K)(I - \tilde{K}^*\tilde{K})^{n-1-j} \Big\| \\
&\quad \leq \Big\| \sum_{j=0}^{p} (I - KK^*)^j (\tilde{K} - K)\tilde{K}^*\tilde{K}(I - \tilde{K}^*\tilde{K})^{n-1-j} \Big\| \qquad (4.113) \\
&\quad\quad + \|(\tilde{K} - K)(I - \tilde{K}^*\tilde{K})^n\| \\
&\quad\quad + \|(I - KK^*)^{p+1}(\tilde{K} - K)(I - \tilde{K}^*\tilde{K})^{n-1-p}\| \\
&\quad\quad + \Big\| \sum_{j=p+1}^{n-1} (I - KK^*)^j KK^*(\tilde{K} - K)(I - \tilde{K}^*\tilde{K})^{n-1-j} \Big\| \\
&\quad \leq \|\tilde{K} - K\| \Big(2 + \sum_{j=0}^{p} (n-j)^{-1} + \sum_{j=p+1}^{n-1} (j+1)^{-1} \Big).
\end{aligned}$$

This together with

$$\sum_{j=0}^{p} (n-j)^{-1} + \sum_{j=p+1}^{n-1} (j+1)^{-1} \leq 2\ln 2, \qquad p := \lfloor \tfrac{n}{2} \rfloor, \qquad (4.114)$$

yields the desired estimate.

Let us now assume that $R \in \mathcal{L}(\mathcal{X}, \mathcal{X})$ with $\|I - R\| \leq c_I < 1$. Using (4.112) with $\tilde{K} = KR$ yields

$$R_\alpha(KR)KR - R_\alpha(K)K = \sum_{j=0}^{n-1}(I - K^*K)^j\Big(K^*K(R - I) + (R^* - I)(R^{-1})^*\tilde{K}^*\tilde{K}\Big)(I - \tilde{K}^*\tilde{K})^{n-1-j}.$$

Applying formula (4.111) twice and using the same sum splitting trick as in estimate (4.113) it follows as above with (4.114) that (4.79) holds.

To show that (4.83) and (4.84) hold we have to use (4.112) with $\tilde{K} = RK$, where $R \in \mathcal{L}(\mathcal{Y},\mathcal{Y})$ with $\|I - R\| \le c_I < 1$, and obtain

$$R_\alpha(RK)RK - R_\alpha(K)K = \sum_{j=0}^{n-1}(I - K^*K)^j K^*\Big((R - I)R^{-1} + (R^* - I)\Big)\tilde{K}(I - \tilde{K}^*\tilde{K})^{n-1-j}.$$

This together with (2.34) yields

$$\|R_\alpha(RK)RK - R_\alpha(K)K\| \le \|I - R\|\,(1 + \|R^{-1}\|)\sum_{j=0}^{n-1}(j+1)^{-\frac{1}{2}}(n-j)^{-\frac{1}{2}}$$

and hence (4.83) holds due to Lemma 2.9.

Applying K from the left side and using formula (4.111) and the sum splitting trick similar as in (4.113), we obtain together with (2.33) (which also holds for $s = 3/2$) and (2.34) the estimate

$$\begin{aligned}
&\|K(R_\alpha(RK)RK - R_\alpha(K)K)\| \\
&\quad\le \|I - R\|\,(1 + \|R^{-1}\|)\Big((n+1)^{-\frac{1}{2}} + (n-q)^{-\frac{1}{2}} \\
&\qquad + \sum_{j=0}^{p}(n-j)^{-\frac{3}{2}} + \sum_{j=p+1}^{n-1}(j+1)^{-1}(n-j)^{-\frac{1}{2}}\Big).
\end{aligned}$$

Choosing $p := \lfloor n/2 \rfloor$, Lemma 2.9 and (4.114) imply that

$$\|K(R_\alpha(RK)RK - R_\alpha(K)K)\| \le c_1\,\|I - R\|\,(n+1)^{-\frac{1}{2}}$$

for some $c_1 > 0$ and thus (4.84) holds.

Finally, we turn to iterated Tikhonov regularization, where the proofs are quite similar to the Landweber case. Using (4.110) with $A_j := \beta_j(K^*K + \beta_j I)^{-1}$ and

$B_j := \beta_j(\tilde{K}^*\tilde{K} + \beta_j I)^{-1}$ we obtain the representation

$$\begin{aligned}
R_\alpha(\tilde{K})\tilde{K} - R_\alpha(K)K
&= \prod_{j=0}^{n} \beta_j(K^*K + \beta_j I)^{-1} - \prod_{j=0}^{n} \beta_j(\tilde{K}^*\tilde{K} + \beta_j I)^{-1} \\
&= \sum_{j=0}^{n} \Big(\prod_{l=0}^{j} \beta_l(K^*K + \beta_l I)^{-1} \Big) \beta_j^{-1} \Big(K^*(\tilde{K} - K) \qquad (4.115) \\
&\qquad + (\tilde{K}^* - K^*)\tilde{K} \Big) \Big(\prod_{l=j}^{n} \beta_l(\tilde{K}^*\tilde{K} + \beta_l I)^{-1} \Big).
\end{aligned}$$

This together with

$$\begin{aligned}
\Big\| (K^*K)^s \prod_{l=m}^{p} \beta_l(K^*K + \beta_l I)^{-1} \Big\| &\le \Big(\sup_{\lambda \ge 0} \lambda \Big(\prod_{l=m}^{p} (1 + \lambda\beta_l^{-1}) \Big)^{-1} \Big)^s \\
&\le \Big(\sum_{l=m}^{p} \beta_l^{-1} \Big)^{-s}, \quad 0 \le s \le 1, \qquad (4.116)
\end{aligned}$$

where $0 \le m \le p$, yields

$$\begin{aligned}
&\|R_\alpha(\tilde{K})\tilde{K} - R_\alpha(K)K\| \\
&\quad \le \|\tilde{K} - K\| \sum_{j=0}^{n} \beta_j^{-1} \Big(\Big(\sum_{l=0}^{j} \beta_l^{-1} \Big)^{-\frac{1}{2}} + \Big(\sum_{l=j}^{n} \beta_l^{-1} \Big)^{-\frac{1}{2}} \Big) \\
&\quad \le c_1 \|\tilde{K} - K\| \alpha_k^{-\frac{1}{2}},
\end{aligned}$$

where $n = n_k$ and c_1 is a positive constant, and hence (4.81) holds. For the last estimate we used the special choice (4.106) for $\{\beta_j\}$ and (4.107).

To obtain an estimate for $\|K(R_\alpha(\tilde{K})\tilde{K} - R_\alpha(K)K)\|$ we proceed as for the Landweber case and use the sum splitting trick as in (4.113). Combining (4.111) and (4.116), and noting that $I - \beta_j(KK^* + \beta_j I)^{-1} = (KK^* + \beta_j I)KK^*$, we obtain that

$$\begin{aligned}
&\|K(R_\alpha(\tilde{K})\tilde{K} - R_\alpha(K)K)\| \\
&\le \|\tilde{K} - K\| \Big(2 + \sum_{j=1}^{p} \beta_{j-1}^{-1} \Big(\sum_{l=j-1}^{n} \beta_l^{-1} \Big)^{-1} + \sum_{j=p+1}^{n} \beta_j^{-1} \Big(\sum_{l=0}^{j} \beta_l^{-1} \Big)^{-1} \\
&\qquad + \sum_{j=0}^{n} \beta_j^{-1} \Big(\sum_{l=0}^{j} \beta_l^{-1} \Big)^{-\frac{1}{2}} \Big(\sum_{l=j}^{n} \beta_l^{-1} \Big)^{-\frac{1}{2}} \Big).
\end{aligned}$$

Since, due to (4.106), the term in brackets is bounded if we choose $p := \lfloor n/2 \rfloor$ for the case $q = 1$ and $p = n$ for the case $q < 1$, this shows that (4.82) holds.

The proofs for the estimates (4.79), (4.83), and (4.84) follow the lines of the Landweber case. Note that for the estimate of (4.84) one needs an appropriate bound like in (4.116) also for $s = 3/2$. This is in fact possible, since

$$\begin{aligned}
&\Big\|(K^*K)^s \prod_{l=m}^{p} \beta_l (K^*K + \beta_l I)^{-1}\Big\| \\
&\quad\leq \Big(\sup_{\lambda \geq 0} \lambda^2 \Big(\prod_{l=m}^{p} (1 + \lambda\beta_l^{-1})\Big)^{-1}\Big)^{\frac{s}{2}} \\
&\quad\leq \Big(\sum_{l=m}^{p-1} \sum_{i=l+1}^{p} \beta_l^{-1}\beta_i^{-1}\Big)^{-\frac{s}{2}}, \quad 0 \leq s \leq 2,
\end{aligned}$$

where $0 \leq m < p$. □

The methodology proposed in this section can also be applied to regularization methods outside the class defined via (4.66). Among those is *regularization by projection*, where the infinite-dimensional linear operator equation is projected to a finite-dimensional subspace $\mathcal{Y}_k$ of the data space $\mathcal{Y}$, where we assume that

$$\mathcal{Y}_1 \subset \mathcal{Y}_2 \subset \mathcal{Y}_3 \subset \ldots \subset \overline{\mathcal{R}(K)}, \qquad \overline{\bigcup_{l \in \mathbb{N}} \mathcal{Y}_l} = \overline{\mathcal{R}(K)}, \tag{4.117}$$

and is solved in a best approximate sense, so that

$$R_\alpha(K) = (Q_l K)^\dagger Q_l = (Q_l K)^\dagger = P_l K^\dagger, \tag{4.118}$$

where Q_l and P_l are the orthogonal projectors onto $\mathcal{Y}_l$ and $\mathcal{X}_l := K^*\mathcal{Y}_l$, respectively. Note that $\|R_\alpha(K)K\| = \|P_l\| = 1$ and that $P_l K^\dagger y \to K^\dagger y$ as $l \to \infty$ (cf. [45, Theorem 3.24]). Moreover, it holds that (cf. [45, Lemma 5.10])

$$\|(I - R_\alpha(K)K)(K^*K)^\mu\| = \|(I - P_l)(K^*K)^\mu\| \leq \tfrac{4}{\pi}\|(I - Q_l)K\|^{2\mu} \tag{4.119}$$

for $\mu \in (0, 1]$ and hence also that

$$\begin{aligned}
\|K(I - R_\alpha(K)K)(K^*K)^\mu\| &\leq \|K(I - P_l)\| \, \|(I - P_l)(K^*K)^\mu\| \\
&\leq \tfrac{4}{\pi}\|(I - Q_l)K\|^{2\mu+1},
\end{aligned} \tag{4.120}$$

where we have used that $Q_l K(I - P_l) = 0$.

Let us assume that the spaces $\mathcal{Y}_l$ have the property that

$$\|(I - Q_l)y\| \leq \tilde{c}_1 h_l^p \|y\|_{\mathcal{Y}_p}$$

for all $y \in \mathcal{Y}_p \subset \mathcal{Y}$, where $p, \tilde{c}_1 > 0$. Such properties are for instance fulfilled for finite element spaces $\mathcal{Y}_l$ and Sobolev spaces $\mathcal{Y}_p$ (cf., e.g., Ciarlet [30]). h_l usually plays the role of a mesh size parameter of the discretization, with $h_l \to 0$ as $l \to \infty$, which is only a suggestive notation, though, and does not exclude meshless discretization methods. If K has the smoothing property that $\mathcal{R}(K) \subset \mathcal{Y}_p$, then we obtain an estimate

$$\|(I - Q_l)K\| \le \tilde{c}_1 \|K\|_{\mathcal{X},\mathcal{Y}_p} h_l^p \,. \tag{4.121}$$

On the other hand, inverse inequalities (cf. [30] in the context of finite elements) yield an estimate for $(Q_l K)^\dagger$ in terms of the mesh size, i.e.,

$$\|(Q_l K)^\dagger\| \le \tilde{c}_2 h_l^{-\tilde{p}} \tag{4.122}$$

for some $\tilde{p}, \tilde{c}_2 > 0$ and all linear operators K satisfying

$$K \in \mathcal{L}(\mathcal{X}, \mathcal{Y}_p)\,, \qquad \|K\|_{\mathcal{X},\mathcal{Y}_p} \le c_s\,, \qquad \overline{\mathcal{R}(K)} = \mathcal{Y}\,. \tag{4.123}$$

With the correspondence

$$\alpha = \alpha_k := h_{l_k}^{2p} \tag{4.124}$$

for the regularization parameter, (4.117) – (4.123) imply that (4.64), (4.69) and (4.70) hold for $\mu \in (0, 1]$, where we additionally restrict the opertors K to those satisfying (4.123). An inspection of the proof of Theorem 4.16, however, shows that the results remain valid, as long as (4.123) holds for all operators $K = F'(x)$ with $x \in \mathcal{B}_{2\rho}(x_0)$.

If the additional condition

$$p = \tilde{p} \tag{4.125}$$

holds, (4.122) and (4.124) imply that R_α defined by (4.118) satisfies (4.78).

To reduce the number of Newton steps, as for the methods in Corollary 4.17, the discretization level l_k should grow as fast as possible still satisfying that $h_{l_k}/h_{l_{k+1}}$ remains bounded so that (4.41) holds.

It turns out that (4.121) – (4.123) and (4.125) are natural conditions in the context of parameter identification and discretization by finite elements. For instance, for the parameter estimation problem of Example 2.14 the conditions are satisfied with $p = \tilde{p} = 2$ if $u(a_0)_s$ is bounded away form zero and if quadratic splines are used for the subspaces $\mathcal{Y}_k$ (cf. [87, 88]).

With these preliminaries, we can prove convergence and convergence rates under Hölder type source conditions if F' satisfies the nonlinearity condition (4.75) with $c_Q = 0$.

Corollary 4.18 *Let Assumption 4.15 (i) and (iii) hold and let* x_k^δ *be defined by the sequence* (4.67)*, where the regularization method* R_α *is defined by* (4.118)*,* (4.124) *and satisfies* (4.121) *and* (4.122) *for all operators as in* (4.123)*. Moreover, we assume that the nonlinearity condition* (4.75) *holds with* $c_Q = 0$*, that* (4.125) *holds, and that* (4.123) *holds for all* $K = F'(x)$ *with* $x \in \mathcal{B}_{2\rho}(x_0)$*.*

If η *and* $\|x_0 - x^\dagger\|$ *are sufficiently small, then, in the noise free case* ($\delta = 0$, $\eta = 0$)*, the sequence* x_k *converges to* $x^\dagger$ *as* $k \to \infty$*. In case of noisy data and with the choice* (4.86)*,* $x_{k_*}^\delta$ *converges to* $x^\dagger$ *as* $\delta \to 0$*.*

If in addition the source condition (2.22) *holds for* $0 < \mu \leq 1/2$ *with* $\|v\|$ *sufficiently small and if* $k_* = k_*(\delta)$ *is chosen according to the stopping rule* (4.87)*, then we obtain the convergence rates*

$$\begin{aligned} \|x_{k_*}^\delta - x^\dagger\| &= O\left(\delta^{\frac{2\mu}{2\mu+1}}\right), \\ \|x_k - x^\dagger\| &= O(\alpha_k^\mu). \end{aligned}$$

Proof. It follows from the considerations above that the results are an immediate consequence of Theorem 4.16 if we can show that (4.80) is satisfied. However, it follows from (4.118), (4.121) – (4.123), and (4.125) that

$$\|KR_\alpha(K)\| \leq \|Q_k K (Q_k K)^\dagger\| + \|(I - Q_k) K (Q_k K)^\dagger\| \leq 1 + \tilde{c}_1 \tilde{c}_2 c_s \,.$$

□

So far we have dealt with a-priori stopping rules. In Section 4.2, we have seen that convergence rates results can be obtained for the iteratively regularized Gauss–Newton method also when stopped via the discrepancy principle (2.2) if $x^\dagger$ satisfies the Hölder type source condition (2.22) with $\mu \leq 1/2$. Results on convergence rates under logarithmic source conditions can be found in [77] for the iteratively regularized Gauss–Newton method, in [37] for the nonlinear Landweber iteration, and, in a very general setting in [79].

For general methods of the form (4.67) convergence rates as in Theorem 4.13 have been proven in [85] for the modified discrepancy principle

$$\max\{\|F(x_{k_*-1}^\delta) - y^\delta\|, \sigma_{k_*}\} \leq \tau\delta < \max\{\|F(x_{k-1}^\delta) - y^\delta\|, \sigma_k\},$$

$1 \leq k < k_*$, where

$$\sigma_k := \|F(x_{k-1}^\delta) + F'(x_{k-1}^\delta)(x_k^\delta - x_{k-1}^\delta) - y^\delta\|,$$

provided that the nonlinearity condition (4.75) and Hölder type source condition (2.22) hold and that $\tau > 1$ is sufficiently large. Moreover, it was shown that $k_*(\delta, y^\delta)$ satisfies the logarithmic bound $O(1 + |\ln\delta|)$ if $\alpha_k \sim q^k$.

In [8], Bauer and Hohage consider a different a posteriori stopping rule and show optimal convergence rates also under Hölder type source conditions (2.22) with $\mu > 1/2$. Acceleration of Newton type methods by preconditioning is proposed and analyzed by Egger in [40]. Burger and Mühlhuber in [23, 24] formulate parameter identification problems for PDEs as constrained minimization problems and consider their solution by sequential quadratic programming.

4.4 Broyden's method for ill-posed problems

As mentioned above, Newton's method exhibits very good convergence properties, namely local quadratic convergence for well-posed problems. However, the need for calculating and inverting the derivative $F'(x)$ in each Newton step means a considerable effort in some applications. Therefore, certain modifications of Newton's method try to reduce this effort by approximating $F'(x)$ and its inverse in such a way that fast convergence can be achieved. A very successful class of such Quasi-Newton methods are those that produce approximates for $F'(x)$ and its inverse by updating with operators of finite rank, usually rank one or two. Among them Broyden's method stands out, which works with rank-one-updates and, under certain conditions, yields superlinear convergence. It was first studied in the finite-dimensional setting in [17, 36] and later on also in general Banach- or Hilbert spaces (see [52, 139]).

In this section we will consider Broyden's method as a regularization method for nonlinear ill-posed problems (1.1). An advantageous feature additional to those mentioned above is that, due to the finite rank also of the inverse update, it suffices to regularize only one linear operator, namely the operator B_0 defining the first quasi-Newton step. For all subsequent steps, we can apply the usual Broyden update. Of course, as in the previously studied iterative methods, also in Broyden's method an appropriate stopping index has to be chosen, due to the noise propagation.

First of all, we will introduce Broyden's method and state some of its important properties for well-posed problems. Then we derive an iterative method for (1.1) based on Broyden's method, where the regularization is achieved by mollifying the data and by stopping the iteration at an index $k = k_*$. Under conditions related to (4.76) we show convergence, convergence rates (under source conditions), and superlinear convergence under a compactness assumption.

We wish to mention that an alternative approach to the one we are using is to apply Broyden's method to a regularized version of (1.1) (see Haber [57]). An advantage of the methodology proposed there is that compactness of the difference between $F'(x^\dagger)$ and its initial approximation B_0 used in the method, as it is required for obtaining superlinear convergence, naturally holds for ill-posed problems. On the other hand an advantage of the method we propose here is, that an appropriate regularization parameter only for a linear problem (see (4.132) below) and not for the fully nonlinear problem (1.1) has to be found.

Broyden's method was first defined in [17] for finite-dimensional regular problems, i.e., problems $F(x) = y$ with $F : \mathbb{R}^n \to \mathbb{R}^n$ having an invertible derivative $F'(x)$, by the following recursion:

$$\begin{aligned} s_k &= -B_k^{-1}(F(x_k) - y)\,, \\ x_{k+1} &= x_k + s_k\,, \\ B_{k+1} &= B_k + \frac{\langle s_k, \cdot \rangle}{\|s_k\|^2}(F(x_{k+1}) - y)\,, \end{aligned} \tag{4.126}$$

as long as B_k is regular and $F(x_k) - y$ and hence also s_k does not vanish. Here x_0 is an initial guess assumed to be sufficiently close to $x^\dagger$, B_0 is some regular and sufficiently good initial approximation to $F'(x^\dagger)$ (e.g., $B_0 := F'(x_0)$), and $x^\dagger$ denotes a solution of $F(x) = y$. Note that B_{k+1} satisfies the so-called secant condition

$$B_{k+1}s_k = F(x_{k+1}) - F(x_k) = \int_0^1 F'(x_k + \theta s_k)s_k \, d\theta \approx F'(x_k)s_k\,,$$

i.e., it approximates the derivative $F'(x_k)$ in the direction s_k. For directions orthogonal to s_k, there is no change in applying B_{k+1} compared to B_k, since

$$(B_{k+1} - B_k)s = 0 \qquad \text{for all} \quad s \perp s_k\,.$$

It is well known, (cf., e.g., [36],) that under a Lipschitz condition on F' (see (4.74)) the operators B_k and B_k^{-1} stay uniformly bounded and Broyden's method converges locally not only linearly

$$\|x_{k+1} - x^\dagger\| \le q\,\|x_k - x^\dagger\| \tag{4.127}$$

with a convergence factor q that can be made the smaller, the closer x_0, B_0 are to $x^\dagger$, $F'(x^\dagger)$, respectively. This would be also the case for the frozen Newton method replacing $F'(x_k)$ by B_0. For Broyden's method, however, convergence occurs even at a superlinear rate, i.e.,

$$\limsup_{k\to\infty} \frac{\|x_{k+1} - x^\dagger\|}{\|x_k - x^\dagger\|} = 0\,. \tag{4.128}$$

Moreover, due to the Sherman-Morrison formula

$$(B + \langle v, \cdot \rangle u)^{-1} = B^{-1} - \frac{\langle v, B^{-1} \cdot \rangle}{1 + \langle v, B^{-1}u \rangle} B^{-1}u\,, \tag{4.129}$$

which holds for regular matrices B as long as $1 + \langle v, B^{-1}u \rangle \neq 0$ (cf., e.g., [36, Lemma 4.2]), B_k is regular for all $k \in \mathbb{N}$ with $H_k = B_k^{-1}$ satisfying the recursion

$$H_{k+1} = H_k - \frac{\langle s_k, H_k \cdot \rangle}{\langle s_k, H_k(F(x_{k+1}) - F(x_k)) \rangle} H_k(F(x_{k+1}) - y),$$

where the denominator does not vanish for all k with $s_k \neq 0$ and x_k sufficiently close to $x^\dagger$, since by the linear convergence (4.127), and uniform boundedness of the operators H_k (cf. [36]),

$$\begin{aligned}
\langle s_k, H_k(F(x_{k+1}) - F(x_k)) \rangle &= \|s_k\|^2 + \langle s_k, H_k(F(x_{k+1}) - y)) \rangle \\
&\geq \|s_k\| \Big(\|x_k - x^\dagger\| - (1 + \|H_k\| \sup_{x \in [x^\dagger, x_{k+1}]} \|F'(x)\|) \|x_{k+1} - x^\dagger\| \Big) \\
&\geq \|s_k\| \Big(1 - q(1 + \|H_k\| \sup_{x \in [x^\dagger, x_{k+1}]} \|F'(x)\|) \Big) \|x_k - x^\dagger\| > 0,
\end{aligned}$$

for $\|x_0 - x^\dagger\|$, $\|B_0 - F'(x^\dagger)\|$, and consequently $q > 0$ sufficiently small.

In the form (4.126), Broyden's method can be immediately defined for regular problems, i.e., continuously invertible $F'(x^\dagger)$ and B_0, also in general, not necessarily finite-dimensional Hilbert spaces ([52, 139]). Moreover, it can be shown that with (4.74) the local linear convergence rate and, if $B_0 - F'(x^\dagger)$ is a compact operator, also the local superlinear convergence rate (4.128) remain valid (cf. [52]).

For ill-posed problems, however, no continuous inverse of $F'(x)$ exists in general. Therefore, the results in [52] cannot be applied directly to ill-posed problems.

It will turn out that nonlinearity conditions like (4.76) are appropriate for analyzing Broyden's method for ill-posed problems. More precisely, the range invariance assumption (4.76) implies that

$$\begin{aligned}
&F'(\tilde{x})^\dagger F'(x) \text{ exists for all } \tilde{x}, x \in \mathcal{B}_{2\rho}(x_0) \text{ and} \\
&\|F'(\tilde{x})^\dagger [F'(x + t(\tilde{x} - x)) - F'(x)](\tilde{x} - x)\| \leq \tilde{L} t \|\tilde{x} - x\|^2, \\
&t \in [0, 1], \quad \tilde{x}, x \in \mathcal{B}_{2\rho}(x_0),
\end{aligned} \tag{4.130}$$

with $\tilde{L} = (1 + 4\rho c_R) c_R$, where we assume that $\mathcal{B}_{2\rho}(x_0) \subset \mathcal{D}(F)$. This is related to the affine covariant Lipschitz condition,

$$\begin{aligned}
&\|F'(\tilde{x})^{-1} [F'(x + t(\tilde{x} - x)) - F'(x)](\tilde{x} - x)\| \leq \tilde{L} t \|\tilde{x} - x\|^2, \\
&t \in [0, 1], \quad \tilde{x}, x \in \mathcal{B}_{2\rho}(x_0),
\end{aligned} \tag{4.131}$$

introduced by Deuflhard, Heindl and Potra in [38, 39] for proving convergence of Newton's method for well-posed problems. Condition (4.131) can be seen as a generalization of the local Lipschitz condition (4.74) on the Fréchet-derivative F'

as it is usual in the proof of quadratic convergence of Newton's method. In [39] it is shown that the assumption that $F'(x)^{-1}$ exists on the domain of F together with (4.131) is sufficient for proving a locally quadratic convergence rate for Newton's method. In [118] Nashed and Chen use a related condition, namely

$$\|F'(x_0)^\sharp(F'(\tilde{x}) - F'(x))\| \leq c\|\tilde{x} - x\|$$

to prove quadratic convergence of a Newton type method

$$x_{k+1} = x_k - F'(x_k)^\sharp F(x_k)$$

with outer inverses $F'(x_k)^\sharp$ of $F'(x_k)$. (Here an outer inverse of a linear operator A is a linear operator $A^\sharp$ with $A^\sharp A A^\sharp = A^\sharp$.)

An important consequence of (4.130) is that for exact data, the Newton step $F'(x_k)^\dagger(y - F(x_k))$ exists, even for an ill-posed problem (1.1). In fact, it can be shown that under condition (4.130) or (4.76), the result in [39] on quadratic convergence of Newton's method remains valid in the ill-posed setting with exact data, see [83, Lemma 2.2]. Thus, in case of noisy data, it suffices to replace y^δ by a mollified version y^δ_α that lies in the range of F, to apply Newton's method, and to stop the iteration at an appropriately chosen index.

We will now show that in this way, one can also adopt Broyden's method for ill-posed problems and preserve superlinear convergence in case of exact data, if some closeness and compactness assumption on B_0 is satisfied. In an asymptotic sense, this fast convergence can also be proven in case of noisy data.

For ill-posed problems one could think of replacing the Broyden step s_k in (4.126) by

$$s^\delta_k = -(B^\delta_k)^\dagger(F(x_k) - y^\delta),$$

where y^δ are nosiy data satisfying (1.2), i.e., $\|y - y^\delta\| \leq \delta$. However, due to the possible non-closedness of the range of B^δ_k, this step will in general not exist for noisy data. Therefore, some regularization technique has to be applied. As mentioned above, we do so by smoothing the data before starting the iteration. For this purpose, we replace y^δ by the mollified version

$$y^\delta_\alpha = F(x_0) + MR_\alpha(M)(y^\delta - F(x_0)), \tag{4.132}$$

where M is a linear bounded operator and R_α is some regularization operator satisfying (4.64), (4.78) and

$$M^\dagger MR_\alpha(M) = R_\alpha(M) = R_\alpha(M)MM^\dagger. \tag{4.133}$$

Note that this condition is fulfilled by any regularization method R_α defined via spectral theory as in (4.66) as well as for regularization by projection (see (4.118)).

If $\alpha = \alpha(\delta)$ is an appropriately chosen regularization parameter (see, e.g., [45]), which we will assume in the following, if (4.76) holds, and if $\mathcal{R}(M) = \mathcal{R}(F'(x))$ (see (4.139) below), then $y_\alpha^\delta \to y$ as $\delta \to 0$.

The *modified Broyden method* is now defined as follows:

$$\begin{aligned} s_k^\delta &= -(B_k^\delta)^\dagger(F(x_k^\delta) - y_\alpha^\delta)\,, \\ x_{k+1}^\delta &= x_k^\delta + s_k^\delta\,, \\ B_{k+1}^\delta &= B_k^\delta + \frac{\langle s_k^\delta, \cdot\rangle}{\|s_k^\delta\|^2}\,(F(x_{k+1}^\delta) - y_\alpha^\delta)\,, \end{aligned} \tag{4.134}$$

with $x_0^\delta := x_0$. The linear bounded operator B_0^δ may depend on δ. In the noise free case, y_α^δ is replaced by y and the superscript δ is omitted, i.e.,

$$\begin{aligned} s_k &= -B_k^\dagger(F(x_k) - y)\,, \\ x_{k+1} &= x_k + s_k\,, \\ B_{k+1} &= B_k + \frac{\langle s_k, \cdot\rangle}{\|s_k\|^2}\,(F(x_{k+1}) - y)\,. \end{aligned} \tag{4.135}$$

In order to guarantee stability of the method, we have to make sure that $(B_k^\delta)^\dagger(F(x_k^\delta) - y_\alpha^\delta)$ can be computed in a stable way. Since B_k^δ differs from B_0^δ only by an operator of finite rank and since we have mollified the data according to (4.132) such that the difference to $F(x_0)$ lies within the range of M, it suffices to have a method for a stable evaluation of $(B_0^\delta)^\dagger(F(x) - F(x_0))$ for any $x \in \mathcal{B}_{2\rho}(x_0)$ and of $(B_0^\delta)^\dagger M R_\alpha(M) z$ for any $z \in \mathcal{Y}$.

If (4.76) holds and if we choose

$$B_0^\delta = B_0 = M = F'(\overline{x}) \quad \text{for some} \quad \overline{x} \in \mathcal{B}_{2\rho}(x_0) \subset \mathcal{D}(F)\,, \tag{4.136}$$

then

$$F'(\overline{x})^\dagger(F(x) - F(x_0)) = F'(\overline{x})^\dagger F'(\overline{x}) \int_0^1 R(x_0 + t(x - x_0), \overline{x})(x - x_0)\,dt\,.$$

In applications, an explicit expression for the integral can usually be derived similarly to condition (4.76) itself (see the examples in [84]). Note that the operator $F'(\overline{x})^\dagger F'(\overline{x})$ is the orthogonal projector onto $\mathcal{N}(F'(\overline{x}))^\perp$. If

$$\mathcal{N}(F'(\overline{x})) = \{0\}\,, \tag{4.137}$$

then $F'(\overline{x})^\dagger F'(\overline{x}) = I$.

Example 4.19 Let us consider the same parameter identification problem as in Example 4.14, namely the identification of c in

$$\begin{aligned} -\Delta u + c\,u &= f\,, && \text{in } \Omega\,,\\ u &= g\,, && \text{on } \partial\Omega\,, \end{aligned}$$

from measurements of u outside $\Omega^* \subset \Omega$. Then we obtain

$$\int_0^1 R(c_0 + t(c - c_0, \bar{c}))(c - c_0)\,dt = u(\bar{c})^{-1} A(\bar{c})(u(c_0) - u(c))$$

provided $u(\bar{c})$ is bounded away from zero, since in this example

$$A(c_0 + t(c - c_0))^{-1}[(c - c_0)u(c_0 + t(c - c_0))] = -\frac{d}{dt} u(c_0 + t(c - c_0))\,.$$

In case u can be measured in all of Ω (i.e., $T = I$ in Example 4.14), condition (4.137) is satisfied.

Assuming that $T = I$, there is also another possibility for choosing M, B_0, and B_0^δ than the one in (4.136), which is more advantageous concerning the convergence of the iterates x_k^δ, namely

$$\begin{aligned} Mh &:= F'(\bar{c})h = -A(\bar{c})^{-1}[h\,u(\bar{c})]\,,\\ B_0 h := -A(\bar{c})^{-1}[h\,u(c^\dagger)]\,, &\qquad B_0^\delta h := -A(\bar{c})^{-1}[h\,u_\alpha^\delta]\,. \end{aligned} \tag{4.138}$$

Here, $u = u(c^\dagger)$ are the exact data and u_α^δ are mollified data as in (4.132) using the measured data u^δ instead of y^δ. Assuming that $u(c^\dagger)$ and u_α^δ are bounded away from zero, we get

$$\begin{aligned} (B_0^\delta)^\dagger M R_\alpha(M) z &= (u_\alpha^\delta)^{-1} u(\bar{c}) R_\alpha(M) z\,,\\ B_0^\dagger M R_\alpha(M) z &= (u(c^\dagger))^{-1} u(\bar{c}) R_\alpha(M) z\,. \end{aligned}$$

The next lemma will be helpful to show that the sequence $\{x_k^\delta\}$ is well defined by (4.134) for all k up to a certain index provided the condition

$$\mathcal{R}(F'(x)) = \mathcal{R}(B_0^\delta) = \mathcal{R}(B_0) = \mathcal{R}(M)\,, \qquad x \in \mathcal{B}_{2\rho}(x_0) \subset \mathcal{D}(F) \tag{4.139}$$

holds. Note that this condition is trivially satisfied for the choice (4.136) if (4.76) holds.

Lemma 4.20 *Let* (4.76) *and* (4.139) *hold, let* $x_k^\delta \in \mathcal{B}_{2\rho}(x_0)$, *and let us assume that* $\mathcal{R}(B_k^\delta) = \mathcal{R}(B_0^\delta)$. *Then* x_{k+1}^δ *is well defined by* (4.134).

If in addition $x_{k+1}^\delta \in \mathcal{B}_{2\rho}(x_0)$ *and*

$$\langle\, s_k^\delta, (B_k^\delta)^\dagger (F(x_{k+1}^\delta) - F(x_k^\delta))\,\rangle \neq 0 \tag{4.140}$$

holds, then B_{k+1}^δ *is well defined by* (4.134), $\mathcal{R}(B_{k+1}^\delta) = \mathcal{R}(B_0^\delta)$, $\mathcal{N}(B_{k+1}^\delta) = \mathcal{N}(B_k^\delta)$, *and the operators* H_{k+1}^δ *and* $\hat{H}_{k+1}^\delta$ *are well defined on* $\mathcal{R}(B_0^\delta)$ *and* $\mathcal{R}(\hat{B}_0^\delta)$, *respectively, and satisfy the recursion formulae*

$$\begin{aligned}
H_{k+1}^\delta &= H_k^\delta - \frac{\langle\, s_k^\delta, H_k^\delta \cdot\,\rangle}{\langle\, s_k^\delta, H_k^\delta (F(x_{k+1}^\delta) - F(x_k^\delta))\,\rangle} H_k^\delta (F(x_{k+1}^\delta) - y_\alpha^\delta)\,,\\
\hat{H}_{k+1}^\delta &= \hat{H}_k^\delta - \frac{\langle\, s_k^\delta, \hat{H}_k^\delta \cdot\,\rangle}{\langle\, s_k^\delta, \hat{H}_k^\delta F'(x^\dagger)^\dagger (F(x_{k+1}^\delta) - F(x_k^\delta))\,\rangle}\\
&\qquad \cdot \hat{H}_k^\delta F'(x^\dagger)^\dagger (F(x_{k+1}^\delta) - y_\alpha^\delta)\,,
\end{aligned}$$

where

$$H_k^\delta := (B_k^\delta)^\dagger\,, \qquad \hat{H}_k^\delta := (\hat{B}_k^\delta)^\dagger\,, \qquad \hat{B}_k^\delta := F'(x^\dagger)^\dagger B_k^\delta\,. \tag{4.141}$$

Here, $x^\dagger$ *denotes a solution of* (1.1), *which is assumed to exist in* $\mathcal{B}_\rho(x_0)$.

Proof. First we show an analogon to the Sherman-Morrison formula (4.129), namely

$$\begin{gathered}
\mathcal{N}(B + \langle\, v, \cdot\,\rangle u) = \mathcal{N}(B)\,, \quad \mathcal{R}(B + \langle\, v, \cdot\,\rangle u) = \mathcal{R}(B)\,, \quad \text{and}\\
(B + \langle\, v, \cdot\,\rangle u)^\dagger = B^\dagger - \frac{\langle\, v, B^\dagger \cdot\,\rangle}{1 + \langle\, v, B^\dagger u\,\rangle} B^\dagger u\,,\\
\text{if } v \in \mathcal{N}(B)^\perp, \quad u \in \mathcal{R}(B)\,, \quad \text{and } 1 + \langle\, v, B^\dagger u\,\rangle \neq 0\,,
\end{gathered} \tag{4.142}$$

for any $B \in L(\mathcal{X}, \mathcal{Y})$, by just checking the Moore–Penrose equations

$$\begin{aligned}
T^\dagger T &= P_{\mathcal{N}(T)^\perp}\,,\\
T T^\dagger &= P_{\overline{\mathcal{R}(T)}}\Big|_{\mathcal{D}(T^\dagger)}\,, \qquad \mathcal{D}(T^\dagger) = \mathcal{R}(T) + \mathcal{R}(T)^\perp\,,\\
T T^\dagger T &= T\,,\\
T^\dagger T T^\dagger &= T^\dagger\,,
\end{aligned} \tag{4.143}$$

that uniquely characterize the generalized inverse $T^\dagger$ of T (cf., e.g., [45, Section 2.1]). Here, $P_{\mathcal{H}}$ denotes the orthogonal projector onto the closed subspace $\mathcal{H}$. Note that the first identity automatically implies the third one. The first identity is satisfied, since

$$(B^\dagger - \frac{\langle\, v, B^\dagger \cdot\,\rangle}{1 + \langle\, v, B^\dagger u\,\rangle} B^\dagger u)(B + \langle\, v, \cdot\,\rangle u) = B^\dagger B + \frac{\langle\, v, [I - B^\dagger B]\cdot\,\rangle}{1 + \langle\, v, B^\dagger u\,\rangle} B^\dagger u = B^\dagger B,$$

which also implies that $\mathcal{N}(B + \langle v, \cdot \rangle u) = \mathcal{N}(B)$. The second identity holds, since

$$
\begin{aligned}
&(B + \langle v, \cdot \rangle u)(B^\dagger - \frac{\langle v, B^\dagger \cdot \rangle}{1 + \langle v, B^\dagger u \rangle} B^\dagger u) \\
&= BB^\dagger + \frac{\langle v, B^\dagger \cdot \rangle}{1 + \langle v, B^\dagger u \rangle}[I - BB^\dagger]u = BB^\dagger ,
\end{aligned}
$$

which also implies that $\mathcal{R}(B + \langle v, \cdot \rangle u) = \mathcal{R}(B)$. Finally, the fourth identity holds, since

$$
\begin{aligned}
&(B^\dagger - \frac{\langle v, B^\dagger \cdot \rangle}{1 + \langle v, B^\dagger u \rangle} B^\dagger u)(B + \langle v, \cdot \rangle u)(B^\dagger - \frac{\langle v, B^\dagger \cdot \rangle}{1 + \langle v, B^\dagger u \rangle} B^\dagger u) \\
&= B^\dagger B(B^\dagger - \frac{\langle v, B^\dagger \cdot \rangle}{1 + \langle v, B^\dagger u \rangle} B^\dagger u) = B^\dagger - \frac{\langle v, B^\dagger \cdot \rangle}{1 + \langle v, B^\dagger u \rangle} B^\dagger u .
\end{aligned}
$$

Thus, (4.142) is shown.

Since (4.76) and (4.139) imply that

$$
\begin{aligned}
F(x_k^\delta) - y_\alpha^\delta &= F(x_k^\delta) - F(x_0) + MR_\alpha(M)(F(x_0) - y^\delta) \\
&\in \mathcal{R}(B_0^\delta) = \mathcal{R}(K) ,
\end{aligned}
\tag{4.144}
$$

where $K := F'(x^\dagger)$, it follows together with $\mathcal{R}(B_k^\delta) = \mathcal{R}(B_0^\delta)$ that x_{k+1}^δ is well defined by (4.134). Moreover, (4.139), (4.141), and (4.143) imply that the operators $\hat{B}_k^\delta$ and $(\hat{B}_k^\delta)^\dagger$ are bounded and satisfy

$$
\begin{aligned}
(\hat{B}_k^\delta)^\dagger &= (B_k^\delta)^\dagger K , & (\hat{B}_k^\delta)^\dagger K^\dagger &= (B_k^\delta)^\dagger , \\
\mathcal{N}(\hat{B}_k^\delta) &= \mathcal{N}(B_k^\delta) , & \mathcal{R}(\hat{B}_k^\delta) &= \mathcal{R}(\hat{B}_0^\delta) = \mathcal{N}(K)^\perp .
\end{aligned}
\tag{4.145}
$$

Let us now assume that $x_{k+1}^\delta \in \mathcal{B}_{2\rho}(x_0)$ and that (4.140) holds. Then $s_k^\delta \neq 0$ and hence B_{k+1}^δ is well defined by (4.134). Since (4.144) now also holds for $k+1$, since $s_k^\delta \in \mathcal{N}(B_k^\delta)^\perp$ and since (4.140) implies that

$$
1 + \|s_k^\delta\|^{-2} \langle s_k^\delta, (B_k^\delta)^\dagger (F(x_{k+1}^\delta) - y_\alpha^\delta) \rangle \neq 0 ,
$$

the remaining assertions follow with (4.134), (4.141), (4.142), and (4.145). □

In the following theorem we will prove a result on local linear convergence of Broyden's method provided that

$$
x_0 - x^\dagger \in \mathcal{N}(B_0^\delta)^\perp = \mathcal{N}(B_0)^\perp \tag{4.146}
$$

holds and that $\|x_0 - x^\dagger\|$ and $\|\hat{B}_0^\delta - K^\dagger K\|$ are sufficiently small.

Due to the propagation of data noise during the iteration, k has to be appropriately stopped. We will show in the next theorem that everything works well until $k_* = k_*(\delta)$ defined by

$$\|s_{k_*}^\delta\| \leq \tau \Delta_\alpha^\delta < \|s_k^\delta\|\,, \qquad 0 \leq k < k_*\,, \tag{4.147}$$

for some $\tau > 0$ sufficiently large, where Δ_α^δ is a computable expression satisfying

$$0 < \Delta_\alpha^\delta \to 0\,, \qquad \tilde{\gamma} \geq \|M^\dagger(y - y_\alpha^\delta)\|(\Delta_\alpha^\delta)^{-1} \to 0 \quad \text{for} \quad \delta \to 0\,, \tag{4.148}$$

where we assume that α is appropriately chosen in dependence of δ and $\tilde{\gamma}$ is some positive constant. If $\tau\Delta_\alpha^\delta < \|s_k^\delta\|$ for all $k \in \mathbb{N}_0$, then $k_* := \infty$. We will give a criterion in the next theorem that guarantees that $k_* < \infty$. In the noise free case, we choose $k_*(0)$ via (4.147) by setting $\Delta_\alpha^\delta = 0$.

If the regularization method R_α used in the definition of y_α^δ (cf. (4.132)) fulfills (4.64), (4.78), and (4.133), then

$$\begin{aligned} \|M^\dagger(y - y_\alpha^\delta)\| &= \|(M^\dagger - R_\alpha(M))(y - F(x_0)) + R_\alpha(M)(y - y^\delta))\| \\ &\leq \|(I - R_\alpha(M)M)M^\dagger(y - F(x_0))\| + O(\delta\alpha^{-\frac{1}{2}}) \\ &= O(\|M^\dagger(y - F(x_0)\| + \delta\alpha^{-\frac{1}{2}})\,. \end{aligned} \tag{4.149}$$

Note that $\|M^\dagger(y - F(x_0))\| < \infty$ if (4.76) and (4.139) hold. Moreover, it holds that $M^\dagger(y - F(x_0)) \to 0$ as $\delta \to 0$.

If R_α also satisfies (4.69) and if the source condition

$$M^\dagger(y - F(x_0)) = (M^*M)^\mu v\,, \qquad v \in \mathcal{X}\,,$$

holds for some $\mu > 0$, then it follows with (4.149) that

$$\|M^\dagger(y - y_\alpha^\delta)\| = o(\delta^{\frac{2\mu}{2\mu+1}})$$

provided α is appropriately chosen. In this case,

$$\Delta_\alpha^\delta = \delta^{\frac{2\mu}{2\mu+1}} \tag{4.150}$$

would satisfy (4.148) and would hence be an appropriate choice in (4.147).

Note that the constant $\tilde{\gamma}$ in (4.148) need not be known explicitely, which is practically relevant, since $\tilde{\gamma}$ has to contain an estimate of the typically unknown norm of the source element v appearing in the source condition. The choice of Δ_α^δ above, however, still relies on the knowledge of the exponent μ or at least on a positive lower bound of μ. If no such bound is available or if $\mu = 0$, then the heuristic estimate

$$\Delta_\alpha^\delta = \delta\alpha^{-\frac{1}{2}}$$

can be used, where α is chosen via an a-posteriori selection criterion (cf. [45, Section 4.3]). Of course, instead of the Hölder type source condition above also logarithmic source conditions (4.68) can be considered.

Theorem 4.21 *Let* (4.76), (4.139), *and* (4.146) *hold and let us assume that* $\|x_0 - x^\dagger\|$ *and* $\|\hat{B}_0^\delta - K^\dagger K\|$ *are sufficiently small and that* $\tau > 0$ *is sufficiently large so that*

$$c(\delta, q) + \tilde{\gamma}\tau^{-1}\|\hat{M}\| \leq \frac{q}{1+q} \tag{4.151}$$

holds for some $q \in (0, 1/3]$, *where*

$$c(\delta, q) := \|\hat{B}_0^\delta - K^\dagger K\| + \tfrac{1}{2}c_R\|x_0 - x^\dagger\|\frac{1+q^2}{(1-q)^2}, \tag{4.152}$$

$\hat{B}_0^\delta := K^\dagger B_0^\delta$, $\hat{M} := K^\dagger M$, $K := F'(x^\dagger)$, $x^\dagger$ *denotes a solution of* (1.1), *which is assumed to exist in* $\mathcal{B}_\rho(x_0)$, c_R *is as in* (4.76), *and* $\tilde{\gamma}$ *is as in* (4.148).

Then the sequence $\{x_k^\delta\}$ *is well defined by* (4.134) *in* $\mathcal{B}_\rho(x^\dagger) \subset \mathcal{B}_{2\rho}(x_0)$ *for all* $0 \leq k \leq k_* + 1$, *where* $k_* = k_*(\delta) < \infty$ *is defined as in* (4.147). *Moreover,*

$$\|x_{k+1}^\delta - x^\dagger\| \leq q\|x_k^\delta - x^\dagger\| \leq q^{k+1}\|x_0 - x^\dagger\| \tag{4.153}$$

and

$$\|\hat{B}_{k+1}^\delta - K^\dagger K\| \leq c(\delta, q) \tag{4.154}$$

hold for all $0 \leq k < k_*$ *and*

$$\|x_{k_*}^\delta - x^\dagger\| = O(\Delta_\alpha^\delta), \qquad \|x_{k_*+1}^\delta - x^\dagger\| = O(\Delta_\alpha^\delta), \tag{4.155}$$

and

$$k_* = O(1 + |\ln \Delta_\alpha^\delta|). \tag{4.156}$$

Proof. We will use an induction argument to show that for all $0 \leq k \leq k_*$ the following assertions hold:

$$x_k^\delta \in \mathcal{B}_{2\rho}(x_0), \ \mathcal{R}(B_k^\delta) = \mathcal{R}(B_0^\delta), \ x_k^\delta - x^\dagger \in \mathcal{N}(B_k^\delta)^\perp = \mathcal{N}(B_0^\delta)^\perp, \tag{4.157}$$

$$\|x_k^\delta - x^\dagger\| \leq q^k\|x_0 - x^\dagger\|, \tag{4.158}$$

$$\|\hat{B}_k^\delta - K^\dagger K\| \leq \|\hat{B}_0^\delta - K^\dagger K\| + \tfrac{1}{2}c_R\|x_0 - x^\dagger\|\frac{(1+q^2)(1-q^k)}{(1-q)^2}. \tag{4.159}$$

Note that, due to (4.139), $\hat{B}_0$ and $\hat{M}$ are bounded.

Since $x_0^\delta = x_0$ and (4.146) holds, the assertions are trivially satisfied for $k = 0$.

Let us now assume that the assertions hold for some $0 \le k < k_*$. Then Lemma 4.20 implies that x_{k+1}^δ is well defined by (4.134).

It follows with (4.134), (4.141), (4.143), (4.145), and $x_k^\delta - x^\dagger \in \mathcal{N}(B_k^\delta)^\perp$ that

$$\begin{aligned} x_{k+1}^\delta - x^\dagger &= x_k^\delta - x^\dagger - \hat{H}_k^\delta K^\dagger (F(x_k^\delta) - y_\alpha^\delta) \\ &= \hat{H}_k^\delta \Big((\hat{B}_k^\delta - K^\dagger K)(x_k^\delta - x^\dagger) \\ &\quad - K^\dagger (F(x_k^\delta) - y - K(x_k^\delta - x^\dagger)) - K^\dagger (y - y_\alpha^\delta) \Big) \end{aligned} \tag{4.160}$$

and that

$$\hat{H}_k^\delta K^\dagger K = \hat{H}_k^\delta \,. \tag{4.161}$$

Therefore,

$$1 = \|\hat{H}_k^\delta \hat{B}_k^\delta\| = \|\hat{H}_k^\delta - \hat{H}_k^\delta (K^\dagger K - \hat{B}_k^\delta)\| \ge \|\hat{H}_k^\delta\| (1 - \|\hat{B}_k^\delta - K^\dagger K\|)\,,$$

which yields the estimate

$$\|\hat{H}_k^\delta\| \le \frac{1}{1 - \|\hat{B}_k^\delta - K^\dagger K\|}\,. \tag{4.162}$$

The image of the Taylor remainder under $K^\dagger$ can be estimated by means of (4.76):

$$\begin{aligned} &\|K^\dagger (F(x_k^\delta) - y - K(x_k^\delta - x^\dagger))\| \\ &\quad = \|K^\dagger \int_0^1 (F'(x^\dagger + t(x_k^\delta - x^\dagger)) - K)(x_k^\delta - x^\dagger)\, dt\| \\ &\quad \le \tfrac{1}{2} c_R \|x_k^\delta - x^\dagger\|^2. \end{aligned} \tag{4.163}$$

For the data error effect one gets with (4.139), (4.141), (4.143), (4.147), and (4.148) the estimate

$$\|K^\dagger (y - y_\alpha^\delta)\| \le \|\hat{M}\|\, \|M^\dagger (y - y_\alpha^\delta)\| \le \tilde{\gamma} \|\hat{M}\| \Delta_\alpha^\delta \le \tilde{\gamma} \tau^{-1} \|\hat{M}\|\, \|s_k^\delta\|. \tag{4.164}$$

Since (4.152), (4.158), and (4.159) imply that

$$\|\hat{B}_k^\delta - K^\dagger K\| + \tfrac{1}{2} c_R \|x_k^\delta - x^\dagger\| \le c(\delta, q)\,, \tag{4.165}$$

we obtain together with (4.134) and (4.160) – (4.164) that

$$(1 - (c(\delta,q) + \tilde{\gamma}\tau^{-1}\|\hat{M}\|))\, \|x_{k+1}^\delta - x^\dagger\| \le (c(\delta,q) + \tilde{\gamma}\tau^{-1}\|\hat{M}\|)\, \|x_k^\delta - x^\dagger\|$$

and hence with (4.151) that

$$\|x_{k+1}^\delta - x^\dagger\| \le q\|x_k^\delta - x^\dagger\|. \tag{4.166}$$

Thus, $x_{k+1}^\delta \in \mathcal{B}_{2\rho}(x_0)$ and (4.158) also holds for $k+1$.

We will now estimate $\langle s_k^\delta, (B_k^\delta)^\dagger(F(x_{k+1}^\delta) - F(x_k^\delta)) \rangle$ from below. Using (4.134), (4.139), (4.141), (4.158), and (4.162) – (4.166), we obtain

$$\begin{aligned}
&\langle s_k^\delta, (B_k^\delta)^\dagger(F(x_{k+1}^\delta) - F(x_k^\delta)) \rangle \\
&\quad = \|s_k^\delta\|^2 + \langle s_k^\delta, \hat{H}_k^\delta K^\dagger(F(x_{k+1}^\delta) - y_\alpha^\delta) \rangle \\
&\quad \ge \|s_k^\delta\| \Big(\|s_k^\delta\| - \|\hat{H}_k^\delta\| (\|x_{k+1}^\delta - x^\dagger\| \\
&\qquad\qquad + \|K^\dagger(F(x_{k+1}^\delta) - y_\alpha^\delta - K(x_{k+1}^\delta - x^\dagger))\| \Big) \\
&\quad \ge \|s_k^\delta\| \Big(\|s_k^\delta\| - \frac{1}{1 - c(\delta,q)} (\|x_{k+1}^\delta - x^\dagger\| + \tfrac{1}{2} c_R \|x_{k+1}^\delta - x^\dagger\|^2 \\
&\qquad\qquad + \tilde{\gamma}\tau^{-1} \|\hat{M}\| \, \|s_k^\delta\|) \Big) \\
&\quad \ge \|s_k^\delta\|^2 \Big(1 - \frac{q + \frac{1}{2} c_R q^{k+2} \|x_0 - x^\dagger\| + (1+q)\tilde{\gamma}\tau^{-1} \|\hat{M}\|}{(1 - c(\delta,q))(1+q)} \Big),
\end{aligned}$$

where we have used the estimate

$$\frac{\|x_{k+1}^\delta - x^\dagger\|}{\|s_k^\delta\|} \le \frac{q}{1-q}. \tag{4.167}$$

Since under the assumption that $q < 1/3$ and that (4.151) holds, the expression in brackets is larger than 0, we may apply Lemma 4.20 to conclude that B_{k+1}^δ is well defined, that $\mathcal{R}(B_{k+1}^\delta) = \mathcal{R}(B_0^\delta)$, and that $\mathcal{N}(B_{k+1}^\delta) = \mathcal{N}(B_k^\delta)$. The last identity together with (4.134) and $x_k^\delta - x^\dagger \in \mathcal{N}(B_k^\delta)^\perp = \mathcal{N}(B_0^\delta)^\perp$ implies that

$$x_{k+1}^\delta - x^\dagger \in \mathcal{N}(B_0^\delta)^\perp = \mathcal{N}(B_k^\delta)^\perp = \mathcal{N}(B_{k+1}^\delta)^\perp.$$

To show (4.159), we define

$$\tilde{B}_{k+1}^\delta := \hat{B}_k^\delta + \frac{\langle s_k^\delta, \cdot \rangle}{\|s_k^\delta\|^2} (K^\dagger K - \hat{B}_k^\delta) s_k^\delta \tag{4.168}$$

for which it obviously holds that

$$(\tilde{B}_{k+1}^\delta - K^\dagger K)s = \begin{cases} (\hat{B}_k^\delta - K^\dagger K)s, & \text{if } s \perp s_k^\delta, \\ 0, & \text{if } s \in \operatorname{span}(s_k^\delta). \end{cases} \tag{4.169}$$

Together with (4.163), which also holds for $k+1$, we now obtain the estimate

$$\begin{aligned}\|\tilde{B}^\delta_{k+1} - \hat{B}^\delta_{k+1}\| &= \Big\|\frac{\langle s^\delta_k, \cdot\rangle}{\|s^\delta_k\|^2}\Big((K^\dagger K - \hat{B}^\delta_k)s^\delta_k - K^\dagger(F(x^\delta_{k+1}) - y^\delta_\alpha)\Big)\Big\| \\ &= \Big\|\frac{\langle s^\delta_k, \cdot\rangle}{\|s^\delta_k\|^2}\Big(K^\dagger(F(x^\delta_k) - F(x^\dagger) - K(x^\delta_k - x^\dagger)) \\ &\qquad - K^\dagger(F(x^\delta_{k+1}) - F(x^\dagger) - K(x^\delta_{k+1} - x^\dagger))\Big)\Big\| \\ &\le \frac{c_R\,(\|x^\delta_k - x^\dagger\|^2 + \|x^\delta_{k+1} - x^\dagger\|^2)}{2\,(\|x^\delta_k - x^\dagger\| - \|x^\delta_{k+1} - x^\dagger\|)}\,. \end{aligned} \tag{4.170}$$

Thus, the difference between $\hat{B}^\delta_k$ and $K^\dagger K$ satisfies the *bounded deterioration condition*

$$\|\hat{B}^\delta_{k+1} - K^\dagger K\| \le \|\hat{B}^\delta_k - K^\dagger K\| + \frac{c_R\,(\|x^\delta_k - x^\dagger\|^2 + \|x^\delta_{k+1} - x^\dagger\|^2)}{2\,(\|x^\delta_k - x^\dagger\| - \|x^\delta_{k+1} - x^\dagger\|)}\,.$$

Note that, due to (4.166) the denominator does not vanish. Together with (4.158), (4.159), and (4.166) we now obtain that

$$\begin{aligned}&\|\hat{B}^\delta_{k+1} - K^\dagger K\| \\ &\quad\le \|\hat{B}_0 - K^\dagger K\| + \tfrac{1}{2}c_R\,\|x_0 - x^\dagger\|\,\frac{(1+q^2)(1-q^k)}{(1-q)^2} \\ &\qquad + \frac{c_R\,\|x^\delta_k - x^\dagger\|^2(1+q^2)}{2\|x^\delta_k - x^\dagger\|\,(1-q)} \\ &\quad\le \|\hat{B}_0 - K^\dagger K\| + \tfrac{1}{2}c_R\,\|x_0 - x^\dagger\|\,\frac{(1+q^2)}{(1-q)^2}(1 - q^k + q^k(1-q))\,,\end{aligned}$$

which implies that (4.159) also holds for $k+1$. This finishes the induction. The estimates (4.153) and (4.154) now follow with (4.152), (4.158), (4.159), and (4.166).

Since (4.134) and (4.166) imply that

$$\|s^\delta_k\| \le (1+q)\,\|x^\delta_k - x^\dagger\|\,, \tag{4.171}$$

we obtain together with (4.147) and (4.158) that

$$0 < \tau\Delta^\delta_\alpha < \|s^\delta_k\| \le (1+q)\,\|x^\delta_k - x^\dagger\| \le (1+q)q^k\,\|x_0 - x^\dagger\|\,.$$

This implies that $k_* < \infty$ and that the estimate (4.156) holds.

Since (4.157) holds for $k = k_*$, also $x^\delta_{k_*+1}$ is well defined. Now (4.154) and (4.160) – (4.165) imply that

$$\|x^\delta_{k_*+1} - x^\dagger\| \le q\|x^\delta_{k_*} - x^\dagger\| + \frac{\tilde{\gamma}\|\hat{M}\|}{1 - c(\delta, q)}\Delta^\delta_\alpha .$$

This together with (4.147) and $\|x^\delta_{k_*} - x^\dagger\| \le \|x^\delta_{k_*+1} - x^\dagger\| + \|s^\delta_{k_*}\|$ immediately yields (4.155). □

Note that for the rate in the theorem above it is necessary that $\|\hat{B}^\delta_0 - K^\dagger K\|$ is sufficiently small for all δ. This is always satisfied at least for sufficiently small δ if $\|\hat{B}_0 - K^\dagger K\|$ is sufficiently small and if $\|\hat{B}^\delta_0 - \hat{B}_0\| \to 0$ as $\delta \to 0$.

If Δ^δ_α is as in (4.150), then (4.156) implies that $k_* = O(1 + |\ln\delta|)$ as for the previously considered Newton type methods.

For the proof of asymptotically superlinear convergence, we need the following lemma, where we estimate the influence of the mollified noisy data. We will assume that $\hat{B}^\delta_0$ and $\hat{B}_0$ satisfy the condition

$$\|\hat{B}^\delta_0 - \hat{B}_0\| \le c_M \|M^\dagger(y - y^\delta_\alpha)\| \tag{4.172}$$

for some positive c_M.

Lemma 4.22 *Let the assumptions of Theorem* 4.21 *hold and assume that* (4.151) *also holds for* $\delta = 0$. *Moreover, assume that* (4.172) *is valid. Then there exists a constant* $c > 0$ *such that for all* $k \le \min\{k_*(\delta), k_*(0)\}$

$$\|x^\delta_k - x_k\| \le c\|M^\dagger(y - y^\delta_\alpha)\| , \tag{4.173}$$

where x^δ_k *and* x_k *are defined by* (4.134) *and* (4.135), *respectively, and* k_* *is chosen by the stopping rule* (4.147).

Proof. First of all, we want to mention that $\{x_k\}$ is also well defined, since an inspection of the proof of Theorem 4.21 shows that the assertions (4.153) and (4.154) remain valid for x_k for all $k < k_*(0)$. Omitting the superscript δ in $\hat{B}^\delta_k$, $\hat{H}^\delta_k$, etc., should reflect the noise free case.

Since (4.145), (4.146), (4.157), and Lemma 4.20 imply that

$$\mathcal{R}(\hat{B}^\delta_k) = \mathcal{N}(K)^\perp = \mathcal{R}(\hat{B}_k) \quad \text{and} \quad \mathcal{N}(\hat{B}^\delta_k) = \mathcal{N}(B^\delta_0) = \mathcal{N}(B_0) = \mathcal{N}(\hat{B}_k) ,$$

we obtain together with (4.143) and (4.161) that

$$\hat{H}_k(\hat{B}^\delta_k - \hat{B}_k)\hat{H}^\delta_k = \hat{H}_k - \hat{H}^\delta_k = \hat{H}^\delta_k(\hat{B}^\delta_k - \hat{B}_k)\hat{H}_k . \tag{4.174}$$

This together with (4.160) yields

$$
\begin{aligned}
x_{k+1}^{\delta} - x_{k+1} &= \hat{H}_k^{\delta}(\hat{B}_k^{\delta} - K^{\dagger}K)(x_k^{\delta} - x^{\dagger}) - \hat{H}_k(\hat{B}_k - K^{\dagger}K)(x_k - x^{\dagger}) \\
&\quad - \hat{H}_k^{\delta}(K^{\dagger}(F(x_k^{\delta}) - y_{\alpha}^{\delta}) - K^{\dagger}K(x_k^{\delta} - x^{\dagger})) \\
&\quad + \hat{H}_k(K^{\dagger}(F(x_k) - y) - K^{\dagger}K(x_k - x^{\dagger})) \\
&= \hat{H}_k(\hat{B}_k^{\delta} - \hat{B}_k)\hat{H}_k^{\delta}(x_k^{\delta} - x^{\dagger}) + \hat{H}_k(\hat{B}_k - K^{\dagger}K)(x_k^{\delta} - x_k) \\
&\quad + \hat{H}_k^{\delta}(\hat{B}_k^{\delta} - \hat{B}_k)\hat{H}_k(K^{\dagger}(F(x_k) - y - K(x_k - x^{\dagger})) \\
&\quad - \hat{H}_k^{\delta}(K^{\dagger}(y - y_{\alpha}^{\delta})) \\
&\quad - \hat{H}_k^{\delta}(K^{\dagger}(F(x_k^{\delta}) - F(x_k) - K(x_k^{\delta} - x_k))\,,
\end{aligned}
$$

where we assume that $k < \min\{k_*(\delta), k_*(0)\}$. Due to (4.154) and (4.162),

$$
\|\hat{H}_k^{(\delta)}\| \leq \frac{1}{1 - c(\delta, q)}\,, \tag{4.175}
$$

where (δ) means that it holds with and without δ. Since (4.158) and (4.165) imply that

$$
\|\hat{B}_k - K^{\dagger}K\| + \tfrac{1}{2}c_R\|x_k^{\delta} - x_k\| \leq c(0, q) + \tfrac{1}{2}c_R q^k\|x_0 - x^{\dagger}\|\,,
$$

we obtain together with (4.163), which also holds for $x^{\dagger}$ replaced by x_k, (4.164), (4.175), and the representation above that

$$
\begin{aligned}
&\|x_{k+1}^{\delta} - x_{k+1}\| \\
&\quad\leq \frac{1}{1 - \tilde{c}(q)}\Big((c(0, q) + \tfrac{1}{2}c_R q^k\|x_0 - x^{\dagger}\|)\,\|x_k^{\delta} - x_k\| \\
&\qquad\qquad + \|\hat{M}\|\,\|M^{\dagger}(y - y_{\alpha}^{\delta})\|\Big) \\
&\qquad + \frac{1}{(1 - \tilde{c}(q))^2}\|\hat{B}_k^{\delta} - \hat{B}_k\|\Big(\|x_k^{\delta} - x^{\dagger}\| + \tfrac{1}{2}c_R q^k\|x_0 - x^{\dagger}\|\,\|x_k - x^{\dagger}\|\Big),
\end{aligned} \tag{4.176}
$$

where

$$
\tilde{c}(q) := \max\{c(\delta, q), c(0, q)\}\,. \tag{4.177}
$$

We will now derive an estimate for $\|\hat{B}_{k+1}^{\delta} - \hat{B}_{k+1}\|$. Using (4.76), (4.134), (4.141), (4.143), (4.157), (4.168), and (4.169), we obtain the representation

$$
\begin{aligned}
\hat{B}_{k+1}^{\delta} - \hat{B}_{k+1} &= (\hat{B}_{k+1}^{\delta} - \tilde{B}_{k+1}^{\delta}) - (\hat{B}_{k+1} - \tilde{B}_{k+1}) \\
&\quad + (\tilde{B}_{k+1}^{\delta} - K^{\dagger}K)(P_{\operatorname{span}(s_k^{\delta})^{\perp}} + P_{\operatorname{span}(s_k^{\delta})}) \\
&\quad - (\tilde{B}_{k+1} - K^{\dagger}K)(P_{\operatorname{span}(s_k)^{\perp}} + P_{\operatorname{span}(s_k)})
\end{aligned}
$$

$$
\begin{aligned}
&= \frac{\langle s_k^\delta, \cdot \rangle}{\|s_k^\delta\|} K^\dagger K (R_k^\delta - I) \frac{s_k^\delta}{\|s_k^\delta\|} - \frac{\langle s_k, \cdot \rangle}{\|s_k\|} K^\dagger K (R_k - I) \frac{s_k}{\|s_k\|} \\
&\quad + (\hat{B}_k^\delta - K^\dagger K) P_{\mathrm{span}(s_k^\delta)^\perp} - (\hat{B}_k - K^\dagger K) P_{\mathrm{span}(s_k)^\perp} \\
&= \left(\frac{\langle s_k^\delta, \cdot \rangle}{\|s_k^\delta\|} - \frac{\langle s_k, \cdot \rangle}{\|s_k\|} \right) K^\dagger K (R_k^\delta - I) \frac{s_k^\delta}{\|s_k^\delta\|} \\
&\quad + \frac{\langle s_k, \cdot \rangle}{\|s_k\|} K^\dagger K (R_k^\delta - R_k) \frac{s_k^\delta}{\|s_k^\delta\|} \\
&\quad + \frac{\langle s_k, \cdot \rangle}{\|s_k\|} K^\dagger K (R_k - I) \left(\frac{s_k^\delta}{\|s_k^\delta\|} - \frac{s_k}{\|s_k\|} \right) \\
&\quad + (\hat{B}_k^\delta - \hat{B}_k) P_{\mathrm{span}(s_k^\delta)^\perp} \\
&\quad + (\hat{B}_k - K^\dagger K)(P_{\mathrm{span}(s_k^\delta)^\perp} - P_{\mathrm{span}(s_k)^\perp}),
\end{aligned}
$$

where we have used the notations

$$
\begin{aligned}
R_k^\delta &:= \int_0^1 R(x_k^\delta + t(x_{k+1}^\delta - x_k^\delta), x^\dagger)\, dt, \\
R_k &:= \int_0^1 R(x_k + t(x_{k+1} - x_k), x^\dagger)\, dt.
\end{aligned}
\tag{4.178}
$$

This together with the estimates

$$
\begin{aligned}
\|P_{\mathrm{span}(s_k^\delta)^\perp} - P_{\mathrm{span}(s_k)^\perp}\| &= \left\| \frac{\langle s_k^\delta, \cdot \rangle}{\|s_k^\delta\|} - \frac{\langle s_k, \cdot \rangle}{\|s_k\|} \right\| = \left\| \frac{s_k^\delta}{\|s_k^\delta\|} - \frac{s_k}{\|s_k\|} \right\| \\
&= \frac{\|(s_k^\delta - s_k)\|s_k^{(\delta)}\| + s_k^{(\delta)}(\|s_k\| - \|s_k^\delta\|)\|}{\|s_k^\delta\|\,\|s_k\|} \\
&\le \frac{2\|s_k^\delta - s_k\|}{\max\{\|s_k^\delta\|, \|s_k\|\}}
\end{aligned}
$$

and

$$
\|I - R^{(\delta)}\| \le \tfrac{1}{2} c_R (\|x_{k+1}^{(\delta)} - x^\dagger\| + \|x_k^{(\delta)} - x^\dagger\|),
$$

$$
\begin{aligned}
\|K^\dagger K (R_k^\delta - R_k)\| &\le \int_0^1 \|K^\dagger K R(x_k + t(x_{k+1} - x_k), x^\dagger) \\
&\qquad (R(x_k^\delta + t(x_{k+1}^\delta - x_k^\delta), x_k + t(x_{k+1} - x_k)) - I)\|\, dt \\
&\le c_R (\|x_{k+1}^\delta - x_{k+1}\| + \|x_k^\delta - x_k\|) \\
&\qquad \cdot \left(\tfrac{1}{2} + \tfrac{1}{3} c_R (\|x_{k+1} - x^\dagger\| + \|x_k - x^\dagger\|) \right),
\end{aligned}
$$

which follow with (4.76) and (4.178), yields

$$\begin{aligned}\|\hat{B}_{k+1}^{\delta} - \hat{B}_{k+1}\| \leq{}& \frac{2\|s_k^{\delta} - s_k\|}{\max\{\|s_k^{\delta}\|, \|s_k\|\}}\Big(\|\hat{B}_k - K^{\dagger}K\| \\ &+ \tfrac{1}{2}c_R(\|x_{k+1}^{\delta} - x^{\dagger}\| + \|x_{k+1} - x^{\dagger}\| \\ &+ \|x_k^{\delta} - x^{\dagger}\| + \|x_k - x^{\dagger}\|)\Big) \\ &+ \|\hat{B}_k^{\delta} - \hat{B}_k\| + c_R(\|x_{k+1}^{\delta} - x_{k+1}\| + \|x_k^{\delta} - x_k\|) \\ &\cdot\Big(\tfrac{1}{2} + \tfrac{1}{3}c_R(\|x_{k+1} - x^{\dagger}\| + \|x_k - x^{\dagger}\|)\Big).\end{aligned} \tag{4.179}$$

Since (4.76), (4.134), (4.141), (4.145), (4.164), (4.174), (4.175), and (4.177) imply that

$$\begin{aligned}\|s_k^{\delta} - s_k\| &= \|\hat{H}_k^{\delta}K^{\dagger}(F(x_k^{\delta}) - y_{\alpha}^{\delta}) - \hat{H}_kK^{\dagger}(F(x_k) - y)\| \\ &\leq \frac{1}{(1-\tilde{c}(q))^2}\|\hat{B}_k^{\delta} - \hat{B}_k\|(1 + \tfrac{1}{2}c_R\|x_k - x^{\dagger}\|)\|x_k - x^{\dagger}\| \\ &\quad+ \frac{1}{1-\tilde{c}(q)}\Big((1 + \tfrac{1}{2}c_R(\|x_k^{\delta} - x^{\dagger}\| + \|x_k - x^{\dagger}\|))\|x_k^{\delta} - x_k\| \\ &\quad+ \|\hat{M}\|\,\|M^{\dagger}(y - y_{\alpha}^{\delta})\|\Big),\end{aligned}$$

we now obtain together with (4.153), (4.152), (4.159), (4.167), and (4.179) that

$$\begin{aligned}&\|x_{k+1}^{(\delta)} - x^{\dagger}\|\,\|\hat{B}_{k+1}^{\delta} - \hat{B}_{k+1}\| \\ &\leq \frac{2q}{(1-q)(1-\tilde{c}(q))}(\tfrac{1}{2}c_R(1+q)q^k\|x_0 - x^{\dagger}\| + c(0,q)) \\ &\quad\cdot\Big(\frac{1 + \tfrac{1}{2}c_Rq^k\|x_0 - x^{\dagger}\|}{1-\tilde{c}(q)}\|\hat{B}_k^{\delta} - \hat{B}_k\|\,\|x_k - x^{\dagger}\| \\ &\quad+ (1 + c_Rq^k\|x_0 - x^{\dagger}\|)\|x_k^{\delta} - x_k\| + \|\hat{M}\|\,\|M^{\dagger}(y - y_{\alpha}^{\delta})\|\Big) \\ &\quad+ c_Rq^{k+1}\|x_0 - x^{\dagger}\|(\tfrac{1}{2} + \tfrac{1}{3}c_R(1+q)q^k\|x_0 - x^{\dagger}\|) \\ &\quad\cdot(\|x_{k+1}^{\delta} - x_{k+1}\| + \|x_k^{\delta} - x_k\|) + q\|x_k^{(\delta)} - x^{\dagger}\|\,\|\hat{B}_k^{\delta} - \hat{B}_k\|.\end{aligned} \tag{4.180}$$

Let us now define

$$\gamma_k := \max\{\tfrac{1}{2}\|x_k^{\delta} - x_k\|,\ \|x_k^{\delta} - x^{\dagger}\|\,\|\hat{B}_k^{\delta} - \hat{B}_k\|,\ \|x_k - x^{\dagger}\|\,\|\hat{B}_k^{\delta} - \hat{B}_k\|\}.$$

Then we obtain with (4.176) and (4.180) for all $k < \min\{k_*(\delta), k_*(0)\}$ that

$$\gamma_{k+1} \leq \max\{c_1, c_2\}\gamma_k + \max\{c_3, c_4\}\|\hat{M}\|\,\|M^{\dagger}(y - y_{\alpha}^{\delta})\|,$$

where

$$
\begin{aligned}
c_1 &:= \frac{c(0,q)+\frac{1}{2}c_R\|x_0-x^\dagger\|}{1-\tilde{c}(q)} + \frac{1+\frac{1}{2}c_R\|x_0-x^\dagger\|}{2(1-\tilde{c}(q))^2}\,,\\
c_2 &:= a\Big(\frac{1+\frac{1}{2}c_R\|x_0-x^\dagger\|}{1-\tilde{c}(q)} + 2(1+c_R\|x_0-x^\dagger\|)\Big) + q + b(1+c_1)\,,\\
c_3 &:= \frac{1}{2(1-\tilde{c}(q))}\,,\\
c_4 &:= a + b\,c_3\,,\\
a &:= \frac{2q}{(1-q)(1-\tilde{c}(q))}(c(0,q)+\tfrac{1}{2}(1+q)c_R\|x_0-x^\dagger\|)\,,\\
b &:= 2qc_R\|x_0-x^\dagger\|(\tfrac{1}{2}+\tfrac{1}{3}(1+q)c_R\|x_0-x^\dagger\|)\,.
\end{aligned}
$$

Note that $\max\{c_1,c_2\} < 1$ if $\|x_0-x^\dagger\|$, $\|\hat{B}_0^\delta - K^\dagger K\|$ and $\|\hat{B}_0 - K^\dagger K\|$ are sufficiently small, which we assume in the following. Note that this also implies that $\tilde{c}(q)$ defined by (4.177) becomes sufficiently small.

Thus, we obtain for all $k \le \min\{k_*(\delta), k_*(0)\}$ that

$$
\|x_k^\delta - x_k\| \le 2\gamma_k \le 2\gamma_0 \max\{c_1,c_2\}^k + \frac{2\max\{c_3,c_4\}}{1-\max\{c_1,c_2\}}\|\hat{M}\|\,\|M^\dagger(y-y_\alpha^\delta)\|\,.
$$

This together with $\gamma_0 = \|x_0-x^\dagger\|\,\|\hat{B}_0^\delta - \hat{B}_0\|$ and (4.172) proves the assertion. □

Essential for the proof of the next theorem is also the following result by Griewank [52]:

Proposition 4.23 *Let $\{C_k\}_{k\in\mathbb{N}}, \{\tilde{C}_k\}_{k\in\mathbb{N}} \subset L(\mathcal{X},\mathcal{X})$ be sequences of positive semidefinite selfadjoint operators, $\{s_k\}_{k\in\mathbb{N}} \subset \mathcal{X}\backslash\{0\}$, $\{\gamma_k\}_{k\in\mathbb{N}} \subset \mathbb{R}$, and c, σ nonnegative constants such that for all $k \in \mathbb{N}$*

$$
\begin{aligned}
&\dim(\mathcal{R}(C_k - C_0)) < \infty\,, && \|C_{k+1} - \tilde{C}_{k+1}\| \le c\gamma_k\,,\\
&\tilde{C}_{k+1}s_k = \sigma\, s_k\,, && \langle v, \tilde{C}_{k+1}v\rangle \le \langle v, C_k v\rangle \quad \textit{if } v \perp s_k\,,\\
&\lambda_*(C_0) \ge \sigma\,, && \sum_{k=0}^{\infty}\gamma_k < \infty\,,
\end{aligned}
$$

where $\lambda_*(C) := \inf\{\|C-T\| \,:\, T \text{ compact}\}$. Then

$$
\limsup_{k\to\infty}\frac{\langle s_k, C_k s_k\rangle}{\|s_k\|^2} \le \lambda_*(C_0)\,.
$$

Proof. See [52, Theorem 3.2]. □

Theorem 4.24 *Let* (4.76), (4.139), (4.146), *and* (4.172) *hold and let us assume that* $\|x_0 - x^\dagger\|$ *and* $\|\hat{B}_0 - K^\dagger K\|$ *are sufficiently small and that* $\tau > 0$ *is sufficiently large. Moreover, let us assume that* $\hat{B}_0 - K^\dagger K$ *is compact. Then*

$$\lim_{\delta \to 0} \frac{\|x^\delta_{k_*} - x^\dagger\|}{\|x^\delta_{k_*-1} - x^\dagger\|} = 0 . \tag{4.181}$$

Moreover, if $k_*(0)$ *is finite, then* $x_{k_*} = x^\dagger$, *otherwise*

$$\limsup_{k \to \infty} \frac{\|x_{k+1} - x^\dagger\|}{\|x_k - x^\dagger\|} = 0 , \tag{4.182}$$

where x^δ_k *and* x_k *are defined by* (4.134) *and* (4.135), *respectively, and* k_* *is chosen by the stopping rule* (4.147). $x^\dagger$ *denotes a solution of* (1.1), *which is assumed to exist in* $\mathcal{B}_\rho(x_0)$.

Proof. First of all, we assume that (4.151) holds for all $\delta \geq 0$ sufficiently small and some $q \in (0, 1/3]$ so that then the results of Theorem 4.21 are applicable. Note that this is possible due to the fact that we assumed that $\|x_0 - x^\dagger\|$ and $\|\hat{B}_0 - K^\dagger K\|$ are sufficiently small and that $\tau > 0$ is sufficiently large and since (4.172) holds.

We will first consider the noise free case. If $k_*(0) = \tilde{k} < \infty$, then it follows with (4.147) that $s^\delta_{\tilde{k}} = 0$ and hence that $x_{\tilde{k}+1} = x_{\tilde{k}}$. Since an inspection of the proof of Theorem 4.21 shows that (4.166) in the noise free case also holds for $k = \tilde{k}$, this shows that $x_{\tilde{k}} = x^\dagger$.

Let us now assume that $k_*(0) = \infty$. We will show that (4.182) holds. By (4.134), (4.141), (4.168), (4.169), (4.170), and Theorem 4.21, the assumptions of Proposition 4.23 are fulfilled for

$$\gamma_k = \|x_k - x^\dagger\| , \qquad \sigma = 0 , \qquad C_k = (\hat{B}_k - K^\dagger K)^*(\hat{B}_k - K^\dagger K) ,$$
$$\tilde{C}_{k+1} = (\tilde{B}_{k+1} - K^\dagger K)^*(\tilde{B}_{k+1} - K^\dagger K) , \quad \text{and} \quad \lambda_*(C_0) = 0 .$$

Thus,

$$\limsup_{k \to \infty} \frac{\|(\hat{B}_k - K^\dagger K)s_k\|}{\|s_k\|} = 0 . \tag{4.183}$$

Now (4.134), (4.141), (4.145), (4.154), (4.163), and (4.175) imply that

$$\begin{aligned}\|x_{k+1} - x^\dagger\| &= \|\hat{H}_k(\hat{B}_k - K^\dagger K)(x_{k+1} - x^\dagger) \\ &\quad + \hat{H}_k(\hat{B}_k - K^\dagger K)(x_k - x_{k+1}) \\ &\quad - \hat{H}_k(K^\dagger(F(x_k) - y - K(x_k - x^\dagger)))\|\end{aligned}$$

$$\leq \frac{1}{1-c(0,q)}\Big(c(0,q)\,\|x_{k+1}-x^\dagger\| + \|(\hat{B}_k - K^\dagger K)s_k\| \\ + \tfrac{1}{2}c_R\,\|x_k - x^\dagger\|^2\Big),$$

where we have used the fact that $(I - \hat{H}_k\hat{B}_k)(x_k - x^\dagger) = 0$, which follows from (4.143) and (4.157).

This estimate together with (4.183) and (4.171) ($\delta = 0$) yields

$$\limsup_{k\to\infty} \frac{\|x_{k+1}-x^\dagger\|}{\|x_k - x^\dagger\|} \leq \frac{1}{1-2c(0,q)} \limsup_{k\to\infty}\Big((1+q)\frac{\|(B_k - K^\dagger K)s_k\|}{\|s_k\|} \\ + \tfrac{1}{2}c_R\,\|x_k - x^\dagger\|\Big) = 0\,.$$

Note that, due to (4.151) and (4.152), $c(0,q) < 1/2$.

Let us now turn to the situation of noisy data. Note that, due to (4.148) and (4.173), x_k^δ depends continuously on δ for all $k \leq \min\{k_*(\delta), k_*(0)\}$.

Let us first assume that $k_*(0) = \tilde{k} < \infty$. Then we now from above that $x_{\tilde{k}} = x^\dagger$. Due to (4.147), this also implies that

$$\liminf_{\delta\to 0} k_*(\delta) \geq \tilde{k}\,. \tag{4.184}$$

Moreover, we obtain together with (4.147), (4.153), and (4.171) that

$$\frac{\|x_{k+1}^\delta - x^\dagger\|}{\|x_k^\delta - x^\dagger\|} \leq \frac{(1+q)\,\|x_{\tilde{k}}^\delta - x_{\tilde{k}}\|}{\tau\Delta_\alpha^\delta}$$

for all $\tilde{k} - 1 \leq k < k_*(\delta)$ and $\delta > 0$ sufficiently small. This together with (4.148) and (4.173) implies the assertion (4.181).

Now we assume that $k_*(0) = \infty$. Then (4.184) implies that $k_*(\delta) \to \infty$. Moreover, (4.147), (4.148), (4.153), (4.173), and (4.171) imply that

$$\begin{aligned}\frac{\|x_{k+1}^\delta - x^\dagger\|}{\|x_k^\delta - x^\dagger\|} &\leq \Big(1 + \frac{\|x_k^\delta - x_k\|}{\|x_k^\delta - x^\dagger\|}\Big)\frac{\|x_{k+1} - x^\dagger\|}{\|x_k - x^\dagger\|} + \frac{\|x_{k+1}^\delta - x_{k+1}\|}{\|x_k^\delta - x^\dagger\|} \\ &\leq \frac{\|x_{k+1} - x^\dagger\|}{\|x_k - x^\dagger\|} + \frac{c(1+q)^2\,\|M^\dagger(y - y_\alpha^\delta)\|}{\tau\Delta_\alpha^\delta}\end{aligned}$$

for all $k < k_*(\delta)$. Thus, assertion (4.181) follows with (4.148) and (4.182). □

Remark 4.25 Note that if we choose M, B_0^δ, and B_0 as in (4.136), then the operator $\hat{B}_0 - K^\dagger K$ will in general not be compact, hence yielding only linear but not superlinear convergence.

If we look at Example 4.19 with the assumptions that $T = I$ and that $u(c^\dagger)$, u_α^δ, and $u(\overline{c})$ are bounded away from zero, and if we choose M, B_0^δ, and B_0 as in (4.138), then the operator $\hat{B}_0 - K^\dagger K$ is compact. This can be shown as follows: using the formula

$$A(c)^{-1}z = A(\tilde{c})^{-1}[z + (\tilde{c} - c)A(c)^{-1}z]\,,$$

one obtains that

$$(\hat{B}_0 - K^\dagger K)h = u(c^\dagger)^{-1}(c^\dagger - \overline{c})A(\overline{c})^{-1}[h\,u(c^\dagger)]\,.$$

The compactness now follows from Sobolev's Embedding Theorem, since $A(\overline{c})^{-1}$ maps $L^2(\Omega)$ into $H^2(\Omega) \cap H_0^1(\Omega)$. Using this argument one can also show that (4.172) holds. Since it was already shown earlier that (4.76) holds and since obviously (4.139) and (4.146) hold, Theorem 4.24 is applicable for this example.

5 Multilevel methods

It is well known from the solution of partial differential equations and optimization problems that multilevel methods have a high numerical efficiency. This is achieved by combining the information on different levels of discretization. Especially multigrid methods play an essential role that use appropriate combinations of smoothing steps on fine grids with coarse grid corrections.

In Section 5.1, we introduce regularization *by* discretization, which provides the basis for the multigrid methods that will be investigated in Section 5.2.

The main result of this chapter concerns the full multigrid method, i.e., a multilevel iteration with multigrid preconditioning, that can be established as an iterative regularization method for nonlinear ill-posed problems.

For combinations of previously presented regularization methods for nonlinear ill-posed problems with discretization on different levels of refinement, we refer to [143] and [37, Section 4].

5.1 Regularization by discretization

Regularization methods for ill-posed operator equations (1.1) are usually defined in an infinite-dimensional setting and have to be discretized for calculating a numerical solution. Since finite-dimensional problems are always well-posed in the sense of stable data dependence one could also think of stabilizing an ill-posed problem *by* discretization.

Projection methods for the regularization of (usually linear) ill-posed problems have been studied and analyzed in a number of publications, whereof we can only cite a short list (cf., e.g., [7, 18, 46, 55, 96, 119, 129, 132, 151]).

An analysis of discretization methods for linear ill-posed problems (cf., e.g., [45, Section 3.3]) shows that, while discretization in the preimage space does not generically converge (cf. a counterexample by Seidman [147]), finite-dimensional projection in the image space always yields a regularization method. Therefore, when considering regularization purely by discretization, we prefer the latter approach. Note that in Section 4.3 we have considered the combination of regularization by projection for linearized problems with a Newton type iteration (see also [86]). Now we apply regularization by image space projection directly to nonlinear problems.

The main purpose of this section is to provide preliminaries and results that will be required for the analysis of multigrid methods in Section 5.2.

A straightforward generalization of regularization by projection to nonlinear problems, $F(x) = y$, is to approximate an exact solution $x^\dagger$ by a solution x_l^δ of the finite-dimensional problem

$$Q_l F(x_l^\delta) = Q_l y^\delta \,, \qquad x_l^\delta \in x_0 + \mathcal{X}_l := x_0 + K_\sharp^* \mathcal{Y}_l \,, \tag{5.1}$$

where Q_l is the orthogonal projector onto the finite-dimensional subspace $\mathcal{Y}_l$,

$$\mathcal{Y}_1 \subset \mathcal{Y}_2 \subset \mathcal{Y}_3 \subset \ldots \subset \overline{\mathcal{R}(K_\sharp)} \,, \qquad \overline{\bigcup_{k\in\mathbb{N}} \mathcal{Y}_k} = \overline{\mathcal{R}(K_\sharp)} \,, \tag{5.2}$$

and $K_\sharp$ is a known bounded linear operator that can be seen as some approximation to $F'(x)$ and that is needed in the numerical computations. As always x_0 is an initial guess for the solution $x^\dagger$. The results of the linear case can be carried over to the nonlinear setting if one of the following two nonlinearity conditions on the operator F is satisfied: either the tangential cone condition (cf. [141])

$$\|F(x) - F(\tilde{x}) - K_\sharp(x - \tilde{x})\| \le c_{tc} \|K_\sharp(x - \tilde{x})\| \,, \quad x, \tilde{x} \in \mathcal{B}_{2\rho}(x_0) \,, \tag{5.3}$$

or the range invariance condition

$$\begin{aligned} &F(x) - F(\tilde{x}) = K_\sharp R(x, \tilde{x})(x - \tilde{x}) \,, \\ &\|R(x, \tilde{x}) - I\| \le c_R \,, \quad x, \tilde{x} \in \mathcal{B}_{2\rho}(x_0) \,, \end{aligned} \tag{5.4}$$

where c_{tc} and c_R are positive constants and $\mathcal{B}_{2\rho}(x_0)$ denotes a closed ball of radius 2ρ around x_0. The last condition means that the range of divided difference operators $DF(x, \tilde{x})$ in

$$F(x) - F(\tilde{x}) = DF(x, \tilde{x})(x - \tilde{x})$$

remains unchanged in $\mathcal{B}_{2\rho}(x_0)$. The divided difference operators $DF(x, \tilde{x})$ are not uniquely defined by this equation. If F is Fréchet-differentiable in $\mathcal{B}_{2\rho}(x_0)$, then one may set

$$DF(x, \tilde{x}) = \int_0^1 F'(\tilde{x} + t(x - \tilde{x}))\, dt \,.$$

If $K_\sharp$ is set to $F'(x_\sharp)$ for some $x_\sharp \in \mathcal{B}_{2\rho}(x_0)$, then (5.3) can usually be verified similarly to (2.4). For the alternative nonlinearity condition (5.4), the previously considered condition (4.76) is sufficient as long as ρ is sufficiently small.

We will here and in the section on multigrid methods concentrate on condition (5.4) and refer to [87, 88, 93] for corresponding results under condition (5.3).

For the convergence analysis of this projection method we need assumptions similar to (4.121) – (4.125):

Assumption 5.1 Let $\{h_l\}_{l\in\mathbb{N}}$ be a real sequence with $h_l \searrow 0$ as $l \to \infty$.

(i) Smoothing and approximation property: let $\mathcal{Y}^p$ be a Banach space such that

$$\mathcal{Y}^p \subset \mathcal{Y}\,, \qquad K_\sharp \in \mathcal{L}(\mathcal{X}, \mathcal{Y}_p)\,, \qquad \|I - Q_l\|_{\mathcal{Y}^p,\mathcal{Y}} \le \tilde{c}_1 h_l^p\,,$$

for some $p, \tilde{c}_1 > 0$.

(ii) Inverse inequality:

$$\underline{\gamma}_l^{-1} \le \tilde{c}_2 h_l^{-p}$$

for some $\tilde{c}_2 > 0$, where

$$\underline{\gamma}_l := \|(Q_l K_\sharp)^\dagger\|^{-1} = \inf\{\,\|K_\sharp^* y\| \;:\; \|y\| = 1 \wedge y \in \mathcal{Y}_l\}\,. \tag{5.5}$$

(iii) Bound on h_l: let

$$\frac{h_l}{h_{l+1}} \le r$$

for some $r > 1$ (compare (4.41)).

It has been mentioned already at the end of Section 4.3 when dealing with projection methods that this assumption is satisfied for parameter identification problems like the one in Example 2.14 if finite elements are used for the spaces $\mathcal{Y}_l$ (cf. [88]).

As usual for nonlinear problems, the initial guess x_0 has to be sufficiently close to $x^\dagger$. Moreover, we assume that the difference $x_0 - x^\dagger$ is orthogonal to the nullspace of $K_\sharp$, i.e.,

$$x_0 - x^\dagger \in \mathcal{N}(K_\sharp)^\perp \tag{5.6}$$

so that it can be expected to be well approximated by functions in $\mathcal{X}_l = K_\sharp^* \mathcal{Y}_l$ (compare (2.9) and (4.146)). This condition together with (5.4) also uniquely selects a solution of (1.1) in $\mathcal{B}_{2\rho}(x_0)$ if c_R is sufficiently small. Moreover, condition (5.4) also guarantees that problem (5.1) has a solution x_l^δ for all l up to some index $\bar{l}^{rd}$ depending on the data noise level. In addition, the approximations x_l^δ satisfy error estimates similar to the linear case (cf. [45, Theorem 3.26]):

Proposition 5.2

(i) Let (5.4) *with* $c_R < 1/2$ *hold and assume that* $x, \tilde{x} \in \mathcal{B}_{2\rho}(x_0)$ *are such that* $F(x) = F(\tilde{x})$ *and* $x - \tilde{x} \in \mathcal{N}(K_\sharp)^\perp$. *Then* $x = \tilde{x}$.

(ii) Let (1.2) *and* (5.4) *with* $c_R < 1$ *hold and let* $\|x_0 - x^\dagger\|$ *be sufficiently small. Then there exist positive constants* c_1^{rd}, c_2^{rd}, *and* c_3^{rd} *independent of* l *and* δ *such that problem* (5.1) *has a solution* x_l^δ *in* $\mathcal{B}_{2\rho}(x_0)$ *for all* $l \le \tilde{l}^{rd}$ *and*

$$\|x_l^\delta - x^\dagger\| \le c_1^{rd}(\|(I - P_l)(x_0 - x^\dagger)\| + \delta\underline{\gamma}_l^{-1}), \tag{5.7}$$

where $\tilde{l}^{rd}$ *is defined by*

$$\tilde{l}^{rd} := \max\{\tilde{l} : c_2^{rd}\|x_0 - x^\dagger\| + c_3^{rd}\delta\underline{\gamma}_l^{-1} \le \rho \textit{ for all } l \le \tilde{l}\}, \tag{5.8}$$

$\underline{\gamma}_l$ *is as in* (5.5)*, and* P_l *is the orthogonal projector onto* $\mathcal{X}_l = K_\sharp^* \mathcal{Y}_l$.

Proof. First we prove (i). It is an immediate consequence of (5.4) that

$$0 = F(x) - F(\tilde{x}) = K_\sharp R(x, \tilde{x})(x - \tilde{x}).$$

This implies that $R(x, \tilde{x})(x - \tilde{x}) \in \mathcal{N}(K_\sharp)$ and hence together with $x - \tilde{x} \in \mathcal{N}(K_\sharp)^\perp$ that

$$R(x, \tilde{x})(x - \tilde{x}) = P_{\mathcal{N}(K_\sharp)} R(x, \tilde{x})(x - \tilde{x}) = P_{\mathcal{N}(K_\sharp)}(R(x, \tilde{x}) - I)(x - \tilde{x}),$$

from which we conclude with (5.4) that

$$\|x - \tilde{x}\| = \|R(x, \tilde{x})^{-1} P_{\mathcal{N}(K_\sharp)}(R(x, \tilde{x}) - I)(x - \tilde{x})\| \le \frac{c_R}{1 - c_R}\|x - \tilde{x}\|.$$

Thus $\|x - \tilde{x}\| = 0$, since $\frac{c_R}{1-c_R} < 1$.

For the proof of (ii) see [87, Theorem 2]. The essential part of the proof deals with existence of a solution of (5.1). This part is quite similar to the verification of solvability of the coarse grid equation in the proof of Proposition 5.4 below. □

Convergence or convergence rates can be achieved if the iteration is appropriately stopped. We suggest to choose the stopping index $l_* = l_*(\delta)$ such that

$$l_* \to \infty \qquad \text{and} \qquad \eta \ge \delta h_{l_*}^{-p} \to 0 \ \text{ as } \ \delta \to 0 \tag{5.9}$$

or via

$$\eta h_{l_*}^{(2\mu+1)p} \le \delta < \eta h_l^{(2\mu+1)p}, \qquad 0 \le l < l_*, \tag{5.10}$$

if the source condition

$$x_0 - x^\dagger = (K_\sharp^* K_\sharp)^\mu v, \qquad \mu > 0, \ v \in \mathcal{N}(K_\sharp)^\perp \tag{5.11}$$

holds. In both cases η is a positive constant.

Assumption 5.1 and (5.6) together with the estimate (5.7) and the stopping criterion (5.9) immediately yield convergence of $x_{l_*}^\delta$ to $x^\dagger$ as $\delta \to 0$.

Moreover, if the source condition (5.11) holds for some $\mu \in (0,1]$ and if the stopping criterion (5.10) is used, then Assumption 5.1, (4.119), and (5.7) imply that

$$\|x_{l_*}^\delta - x^\dagger\| = O(\delta^{\frac{2\mu}{2\mu+1}}).$$

Note that $l_* \leq \tilde{l}^{rd}$ (cf. (5.8)) if η is sufficiently small.

See [87] for results on optimal convergence rates where instead of (5.9) and (5.10) the discrepancy principle (2.2) (with k replaced by l) is used.

5.2 Multigrid methods

Multigrid methods (MGM) are well known to be extremely efficient solvers for large scale systems of equations resulting from partial differential equations and integral equations of the second kind and have been extensively studied in the last years (see, e.g., [13, 14, 56, 59]). A well-known application of multigrid operators is the full multigrid method (FMGM), i.e., the simple nested iteration scheme starting at the coarsest level and consisting of multigrid preconditioned fixed point steps on gradually refining levels.

For ill-posed problems the behaviour of MGM is not so well understood, yet. The main problem lies in the definition of smoothing operators. The crucial effect of smoothing out the higher frequency part of the error, that the usual iterative schemes such as Jacobi, Gauss–Seidel, or SOR show in discretizations of PDEs or second kind integral equations, gets lost for first kind integral equations, due to their adverse eigensystem structure: here high frequencies correspond to small singular values of the forward operator and are therefore strongly amplified by inversion.

An analysis of multigrid methods for (Tikhonov) regularized ill-posed problems, with possibly small but positive regularization parameter α and a discretization in preimage space can be found in [68, 136] (see also [58]). We here concentrate more on the aspect of regularization solely *by discretization*, and therefore apply projection in image space for the reasons mentioned above, still including the possibility of additional regularization (see [88]).

The simplest case of a two grid method is sketched in Figure 5.1 and can be used to recursively define a multigrid method by replacing the exact solution on the coarser level again by a two grid step.

The nonlinear multigrid method (cf. [59], see also [14] for a different variant) for the approximate solution of the projected equation (1.1) with noisy data y^δ, i.e.,

$$Q_l F(x) = Q_l y^\delta =: f_l \in \mathcal{Y}_l$$

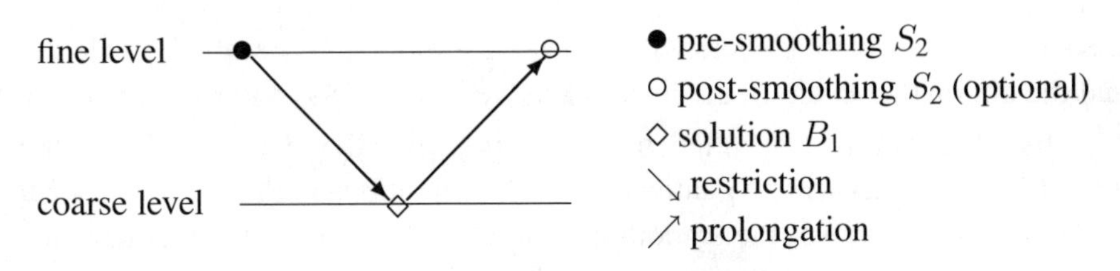

Figure 5.1 The two grid method B_2

in $x_0 + \mathcal{X}_l = x_0 + K_\sharp^* \mathcal{Y}_l$, with $K_\sharp$ as in (5.3) or (5.4), can be inductively defined as follows:

Algorithm 5.3 (Nonlinear Multigrid Method, NMGM)
Let $x_{\text{init}} \in (x_0 + \mathcal{X}_l) \cap \mathcal{B}_\rho(x_0)$, $f_l \in \mathcal{Y}_l$.

(i) Set $\text{NMGM}(1, x_{\text{init}}, f_l) := \Phi_1(x_{\text{init}}, f_l) \in x_0 + \mathcal{X}_1$.

(ii) Assume that $\text{NMGM}(l-1, s, d)$ has been defined for all $s \in (x_0 + \mathcal{X}_{l-1}) \cap \mathcal{B}_\rho(x_0)$ and $d \in \mathcal{Y}_{l-1}$ and that it maps into $x_0 + \mathcal{X}_{l-1}$. Moreover, let $\tilde{x}_{l-1} \in (x_0 + \mathcal{X}_{l-1}) \cap \mathcal{B}_\rho(x_0)$ and $\tilde{f}_{l-1} := Q_{l-1} F(\tilde{x}_{l-1})$ be given.

(iii) Pre-smoothing: Set $x_{\text{smo}} := S_l(x_{\text{init}}, f_l)$.

Coarse grid correction: Set $d_{l-1} := Q_{l-1}(F(x_{\text{smo}}) - f_l)$
Choose $\lambda_{l-1} \in \mathbb{R}^+$
Set $\tilde{d}_{l-1} := \tilde{f}_{l-1} - \lambda_{l-1} d_{l-1}$
Set $s_{l-1}^{(0)} := \tilde{x}_{l-1}$
For $i = 1, \ldots, m$:
Set $s_{l-1}^{(i)} = \text{NMGM}(l-1, s_{l-1}^{(i-1)}, \tilde{d}_{l-1})$

Set $\text{NMGM}(l, x_{\text{init}}, f_l) := x_{\text{smo}} + \lambda_{l-1}^{-1}(s_{l-1}^{(m)} - \tilde{x}_{l-1})$.

Here $\Phi_1(x_{\text{init}}, f_l)$ is a (typically iterative) solution method on the coarsest grid that is supposed to approximate a solution of $Q_1 F(x) = Q_1 f_l$ in $x_0 + \mathcal{X}_1$ in a sense to be specified below (see (5.22)).

The damping parameter λ_{l-1} must in principle be chosen at each level such that certain closeness requirements (see (5.20) below) are satisfied. The auxiliary values $\tilde{x}_{l-1}$, $\tilde{f}_{l-1}$, that are used for constructing starting points for the coarse grid iteration, are here preliminarily assumed to be a-priori fixed. In the nested iteration (see Algorithm 5.5 below), where the operators NMGM are used, they are defined in the course of the iteration.

To simplify notation, we use a cycle regime m that is constant over the discretization levels. Of course, also a level dependent cycle regime $m = m(l)$ can be used, e.g., the alternating cycle with $m(l) = 1 + \mathrm{mod}(l,2)$. For the V-cycle $m \equiv 1$ and the alternating cycle in the linear case we refer to [89] and [88], respectively. Note that Algorithm 5.3 is well defined only if $\mathrm{NMGM}(l-1, s_{l-1}^{(i-1)}, \tilde{d}_{l-1})$ remains within $\mathcal{B}_\rho(x_0)$. This will be guaranteed by the conditions in Proposition 5.4 below.

As mentioned above, the smoothing operators S_l have to be carefully chosen. In [94], King shows that an appropriate choice in the linear case, $F(x) = K_\sharp x$, is the operator

$$S_l^{\mathrm{lin}} := \xi_l K_\sharp^* (I - Q_{l-1}) Q_l \tag{5.12}$$

with $\xi_l \approx \|(Q_l K_\sharp)^\dagger\|^2$, more precisely,

$$\xi_l \overline{\gamma}_{l-1}^2 \leq 1 \qquad \text{and} \qquad \xi_l \underline{\gamma}_l^2 \geq \underline{c} > 0\,, \tag{5.13}$$

where $\underline{\gamma}_l$ is as in (5.5) and $\overline{\gamma}_l$ is defined by

$$\overline{\gamma}_l := \|(I - Q_l) K_\sharp\|\,. \tag{5.14}$$

Note that

$$\begin{aligned}
\underline{\gamma}_l &= \inf\{\|Q_l K_\sharp v\| \,:\, \|v\| = 1 \wedge v \in \mathcal{X}_l\} \\
&\leq \inf\{\|Q_l K_\sharp v\| \,:\, \|v\| = 1 \wedge v \in \mathcal{X}_{l-1}^\perp \cap \mathcal{X}_l\} \\
&= \inf\{\|Q_l (I - Q_{l-1}) K_\sharp v\| \,:\, \|v\| = 1 \wedge v \in \mathcal{X}_{l-1}^\perp \cap \mathcal{X}_l\} \leq \overline{\gamma}_{l-1}\,,
\end{aligned}$$

since

$$\mathcal{N}(Q_{l-1} K_\sharp) = \mathcal{R}(K_\sharp^* Q_{l-1}) = \mathcal{X}_{l-1}^\perp \tag{5.15}$$

and hence $\underline{c} \leq 1$ has to hold in (5.13). If Assumption 5.1 holds, an appropiate choice for ξ_l is given by

$$\xi_l \sim h_l^{-2p}.$$

We mention in passing that to achieve optimal convergence factors one should choose ξ_l as large as possible subject to the constraint (5.13). Thus, for computational purposes, it sometimes pays off to precompute (an approximation of) $\overline{\gamma}_{l-1}$ and set

$$\xi_l = \overline{\gamma}_{l-1}^{-2}\,.$$

The operator S_l^{lin} acts as a stabilized rough approximation of the inverse of $Q_l K_\sharp$ on the *high frequency part* $(I - Q_{l-1})\mathcal{Y}_l$ of $\mathcal{Y}_l$, while just removing the *low frequency part* $Q_{l-1}\mathcal{Y}_l = \mathcal{Y}_{l-1}$, leaving its treatment to the coarse grid correction.

The operator $I - S_l^{\mathrm{lin}} K_\sharp$ is symmetric with its spectrum contained in $(0,1)$: note that $S_l^{\mathrm{lin}} K_\sharp = \xi_l K_\sharp^* (Q_l - Q_{l-1}) K_\sharp$, where we use the fact that in nested spaces $\mathcal{Y}_l$ $Q_{l-1} Q_l = Q_{l-1} = Q_l Q_{l-1}$ holds. Moreover, (5.13) and (5.14) imply that

$$\langle I - S_l^{\mathrm{lin}} K_\sharp v, v\rangle = \|v\|^2 - \xi_l \|Q_l (I - Q_{l-1}) K_\sharp v\|^2 \geq (1 - \xi_l \overline{\gamma}_{l-1}^2)\|v\|^2 \geq 0$$

for all $v \in \mathcal{X}$ and hence

$$\|I - S_l^{\text{lin}} K_\sharp\| \le 1 \qquad \text{and} \qquad \|S_l^{\text{lin}} K_\sharp\| \le 1\,. \tag{5.16}$$

In addition, it can be shown as in [94, Lemma 2.5] that $I - S_l^{\text{lin}} K_\sharp$ is a contraction on $\mathcal{X}_{l-1}^\perp \cap \mathcal{X}_l$: using $Q_{l-1} K_\sharp (I - P_{l-1}) = 0$ (cf. (5.15)), we obtain that

$$\begin{aligned}(I - S_l^{\text{lin}} K_\sharp) P_l (I - P_{l-1}) &= (I - S_l^{\text{lin}} K_\sharp)(I - P_{l-1}) P_l \\ &= (I - \xi_l K_\sharp^* Q_l K_\sharp) P_l (I - P_{l-1})\end{aligned}$$

and for all $v \in \mathcal{X}_l$ that

$$\langle (I - \xi_l K_\sharp^* Q_l K_\sharp) v, v \rangle = \|v\|^2 - \xi_l \|Q_l K_\sharp v\|^2 \le (1 - \xi_l \underline{\gamma}_l^2) \|v\|^2.$$

This together with (5.13) implies that

$$\|(I - S_l^{\text{lin}} K_\sharp) P_l (I - P_{l-1})\| \le \sqrt{1 - \underline{c}}\,. \tag{5.17}$$

Generalizing the smoothing operator (5.12) to the nonlinear situation, we set

$$\begin{aligned} S_l(x_{\text{init}}, f_l) \;&:=\; x_{\text{init}} - \xi_l K_\sharp^* (I - Q_{l-1})(Q_l F(x_{\text{init}}) - f_l)\,, \\ & \quad\; x_{\text{init}} \in \mathcal{X}_l,\; f_l \in \mathcal{Y}_l\,, \end{aligned} \tag{5.18}$$

with ξ_l chosen according to (5.13).

The uniform contraction property of $I - S_l^{\text{lin}} K_\sharp$ on the high frequency part of level l, in combination with the coarse grid correction, enables us to prove a uniform contraction factor estimate for $I - \text{NMGM}(l, x_{\text{init}}, Q_l F)$.

Proposition 5.4 *Let* (5.4) *with sufficiently small* $c_R < 1$ *hold and let* NMGM *be as in Algorithm 5.3 with* S_l *defined as in* (5.18), *where* ξ_l *satisfies* (5.13). *Moreover, we assume that the cycle regime* m *is chosen such that*

$$\underline{c} m > 1 \tag{5.19}$$

and that the sequences $\{\lambda_l\}$, $\{\tilde{x}_l\}$ *are such that*

$$4 \frac{(1 + c_R)^2}{1 - c_R} \lambda_l \tilde{\rho} + \|\tilde{x}_l - x_0\| \le \tilde{\rho} \tag{5.20}$$

for all $l \in \mathbb{N}$, *where* $\tilde{\rho} > 0$ *satisfies*

$$(3 + 2c_R)\tilde{\rho} \le 2\rho\,. \tag{5.21}$$

In addition, we assume that

$$\|\Phi_1(x_{\text{init}}, Q_1 F(x)) - x\| \le \varepsilon \|x_{\text{init}} - x\| \tag{5.22}$$

for all $x_{\text{init}}, x \in (x_0 + \mathcal{X}_1) \cap \mathcal{B}_{\tilde{\rho}}(x_0)$, *where*

$$\varepsilon := (\underline{c}m)^{-\frac{1}{2(m-1)}} \tag{5.23}$$

and $\underline{c}$ *is as in* (5.13).

Then NMGM *has the level independent contraction factor* ε, *i.e.,*

$$\| \text{NMGM}(l, x_{\text{init}}, Q_l F(x)) - x\| \le \varepsilon \|x_{\text{init}} - x\| \tag{5.24}$$

for all $l \in \mathbb{N}$ *and all* $x_{\text{init}}, x \in (x_0 + \mathcal{X}_l) \cap \mathcal{B}_{\tilde{\rho}}(x_0)$.

Proof. The proof is done by induction. For $l = 1$, (5.24) trivially holds by (5.22) and Algorithm 5.3 (i). Let us now assume that (5.24) holds with l replaced by $l-1$ and ε as in (5.23), and let $x_{\text{init}}, x \in (x_0 + \mathcal{X}_l) \cap \mathcal{B}_{\tilde{\rho}}(x_0)$ be arbitrary but fixed.

Denoting by

$$x_{\text{smo}} := S_l(x_{\text{init}}, Q_l F(x))$$

the result of the smoothing step, it follows with Algorithm 5.3 (iii) and (5.18) that $\text{NMGM}(l, x_{\text{init}}, Q_l F(x)) \in x_0 + \mathcal{X}_l$ and that we can decompose the error into its relatively high and relatively low frequent contributions as follows

$$\begin{aligned} \text{NMGM}(l, x_{\text{init}}, Q_l F(x)) - x = &\underbrace{(I - P_{l-1})(x_{\text{smo}} - x)}_{\in \mathcal{X}_{l-1}^{\perp}} \\ &+ \underbrace{P_{l-1}(x_{\text{smo}} - x) + \lambda_{l-1}^{-1}(s_{l-1}^{(m)} - \tilde{x}_{l-1})}_{\in \mathcal{X}_{l-1}} . \end{aligned} \tag{5.25}$$

Due to (5.4), (5.12), and (5.18), the difference between x_{smo} and x can be written as

$$\begin{aligned} x_{\text{smo}} - x &= x_{\text{init}} - x - \xi_l K_\sharp^* [I - Q_{l-1}] Q_l (F(x_{\text{init}}) - F(x)) \\ &= (I - S_l^{\text{lin}} K_\sharp)(x_{\text{init}} - x) + S_l^{\text{lin}} K_\sharp (I - R(x_{\text{init}}, x))(x_{\text{init}} - x) , \end{aligned}$$

and hence, using the properties (5.16) and (5.17) of S_l^{lin}, we get the estimates

$$\|x_{\text{smo}} - x\| \le (1 + c_R) \|x_{\text{init}} - x\| , \tag{5.26}$$

$$\|(I - P_{l-1})(x_{\text{smo}} - x)\| \le (\sqrt{1 - \underline{c}} + c_R) \|x_{\text{init}} - x\| . \tag{5.27}$$

Moreover, we obtain together with (5.21) that $\|x_{\text{smo}} - x_0\| \le 2\rho$ and hence (5.4) is applicable to x_{smo}.

To make the induction hypothesis applicable, we now show that the coarse grid equation

$$Q_{l-1}F(s) = \tilde{d}_{l-1} \tag{5.28}$$

with

$$\tilde{d}_{l-1} = Q_{l-1}(F(\tilde{x}_{l-1}) - \lambda_{l-1}(F(x_{\text{smo}}) - F(x))), \tag{5.29}$$

has a solution $\overline{s}_{l-1} \in (x_0 + \mathcal{X}_{l-1}) \cap \mathcal{B}_{\tilde{\rho}}(x_0)$ (that is supposed to be approximated by $s_{l-1}^{(i)}$ as given in Algorithm 5.3 (iii)): consider the sequence

$$s^0 := \tilde{x}_{l-1}, \qquad s^{j+1} = s^j - (Q_{l-1}K_\sharp)^\dagger(Q_{l-1}F(s^j) - \tilde{d}_{l-1}), \tag{5.30}$$

which obviously remains in $x_0 + \mathcal{X}_{l-1}$. Then we obtain together with (5.4), (5.26), and (5.29) that

$$\begin{aligned}
\|s^1 - s^0\| &\le \|(Q_{l-1}K_\sharp)^\dagger Q_{l-1}(F(\tilde{x}_{l-1}) - \tilde{d}_{l-1}\| \\
&\le \lambda_{l-1}\|(Q_{l-1}K_\sharp)^\dagger Q_{l-1}(F(x_{\text{smo}}) - F(x)\| \\
&\le \lambda_{l-1}\|R(x_{\text{smo}}, x)(x_{\text{smo}} - x)\| \\
&\le \lambda_{l-1}(1 + c_R)^2\|x_{\text{init}} - x\| \le 2\lambda_{l-1}(1 + c_R)^2\tilde{\rho},
\end{aligned}$$

which together with (5.20) implies that $s^0, s^1 \in \mathcal{B}_{\tilde{\rho}}(x_0) \subset \mathcal{B}_{2\rho}(x_0)$. It now follows with induction, (5.4), (5.20), the estimate

$$\begin{aligned}
\|s^{j+1} - s^j\| &= \|s^j - s^{j-1} - (Q_{l-1}K_\sharp)^\dagger Q_{l-1}(F(s^j) - F(s^{j-1}))\| \\
&= \|(Q_{l-1}K_\sharp)^\dagger Q_{l-1}K_\sharp(I - R(s^j, s^{j-1}))(s^j - s^{j-1})\| \\
&\le c_R\|s^j - s^{j-1}\| \le 2c_R^j\lambda_{l-1}(1 + c_R)^2\tilde{\rho},
\end{aligned}$$

and the triangle inequality that the sequence $\{s^j\}$ remains in $\mathcal{B}_{\tilde{\rho}}(x_0)$. Moreover, it follows that $\{s^j\}$ is a Cauchy sequence and, therefore, converges to some $\overline{s}_{l-1} \in (x_0 + \mathcal{X}_{l-1}) \cap \mathcal{B}_{\tilde{\rho}}(x_0)$, which, by taking limits $j \to \infty$ in (5.30), solves (5.28), i.e.,

$$Q_{l-1}(\lambda_{l-1}^{-1}(F(\overline{s}_{l-1}) - F(\tilde{x}_{l-1})) + F(x_{\text{smo}}) - F(x)) = 0. \tag{5.31}$$

We can now apply the induction hypothesis to obtain that

$$\|s_{l-1}^{(1)} - \overline{s}_{l-1}\| \le \varepsilon\|\tilde{x}_{l-1} - \overline{s}_{l-1}\|.$$

Thus, it follows together with the estimate

$$\|s_{l-1}^{(1)} - x_0\| \le \|\tilde{x}_{l-1} - x_0\| + 2\|\tilde{x}_{l-1} - \overline{s}_{l-1}\|$$

and (5.20) that $s_{l-1}^{(1)} \in \mathcal{B}_{\tilde{\rho}}(x_0)$. By induction we obtain that $s_{l-1}^{(i)}$ remains in $\mathcal{B}_{\tilde{\rho}}(x_0)$ and that

$$\|s_{l-1}^{(m)} - \bar{s}_{l-1}\| \le \varepsilon^m \|\tilde{x}_{l-1} - \bar{s}_{l-1}\| . \tag{5.32}$$

Due to the nonlinearity of the problem, there is a non-vanishing discrepancy on the coarse grid, namely

$$\partial s := \lambda_{l-1}^{-1}(\bar{s}_{l-1} - \tilde{x}_{l-1}) + P_{l-1}(x_{\text{smo}} - x) . \tag{5.33}$$

(Note that in the linear case $\partial s = 0$.) This ∂s also enters into the total error, and can be estimated as follows: by (5.4), and applying $(Q_{l-1}K_\sharp)^\dagger$ to (5.31), we can write

$$0 = P_{l-1}(\lambda_{l-1}^{-1} R(\bar{s}_{l-1}, \tilde{x}_{l-1})(\bar{s}_{l-1} - \tilde{x}_{l-1}) + R(x_{\text{smo}}, x)(x_{\text{smo}} - x))$$

and, therefore,

$$\begin{aligned} \partial s &= P_{l-1}\partial s = P_{l-1}R(\bar{s}_{l-1}, \tilde{x}_{l-1})\partial s + P_{l-1}(I - R(\bar{s}_{l-1}, \tilde{x}_{l-1}))\partial s \\ &= -P_{l-1}R(x_{\text{smo}}, x)(x_{\text{smo}} - x) + P_{l-1}R(\bar{s}_{l-1}, \tilde{x}_{l-1})(x_{\text{smo}} - x) \\ &\quad + P_{l-1}(I - R(\bar{s}_{l-1}, \tilde{x}_{l-1}))\partial s , \end{aligned}$$

which yields

$$\begin{aligned} \|\partial s\| &= \|(I + P_{l-1}(R(\bar{s}_{l-1}, \tilde{x}_{l-1}) - I))^{-1} P_{l-1} \\ &\qquad (R(\bar{s}_{l-1}, \tilde{x}_{l-1}) - R(x_{\text{smo}}, x))(x_{\text{smo}} - x)\| \\ &\le \frac{2c_R}{1 - c_R} \|x_{\text{smo}} - x\| \end{aligned}$$

and, by the triangle inequality, also

$$\begin{aligned} \lambda_{l-1}^{-1} \|\tilde{x}_{l-1} - \bar{s}_{l-1}\| &= \|\partial s - P_{l-1}(x_{\text{smo}} - x)\| \\ &\le \frac{2c_R}{1 - c_R} \|x_{\text{smo}} - x\| + \|P_{l-1}(x_{\text{smo}} - x)\| . \end{aligned}$$

Together with (5.25), (5.32), and (5.33) we finally obtain the estimate

$$\begin{aligned} &\| \operatorname{NMGM}(l, x_{\text{init}}, Q_l F(x)) - x\|^2 \\ &\quad \le \varphi(\|[(I - P_{l-1})(x_{\text{smo}} - x)\|, \|x_{\text{smo}} - x\|) \end{aligned} \tag{5.34}$$

with

$$\varphi(a,b) := a^2c_1 + b^2c_2 + b(b^2-a^2)^{\frac{1}{2}}c_3\,, \tag{5.35}$$

$$\begin{aligned} c_1 &:= 1-\varepsilon^{2m}, \\ c_2 &:= \varepsilon^{2m} + \frac{4c_R^2}{(1-c_R)^2}(1+\varepsilon^m)^2, \\ c_3 &:= \frac{4c_R}{1-c_R}\varepsilon^m(1+\varepsilon^m)\,. \end{aligned} \tag{5.36}$$

Obviously φ is monotonically increasing with respect to b. Since

$$\frac{\partial\varphi}{\partial a}(a,b) = a(2c_1 - b(b^2-a^2)^{-\frac{1}{2}}c_3)\,,$$

φ is monotonically increasing for all $0 \le a \le \overline{b}$ if

$$4c_1^2a^2 \le \overline{b}^2(4c_1^2 - c_3^2)\,.$$

According to (5.26), (5.27), and (5.34) – (5.36), this means that

$$\begin{aligned} &\|\,\mathrm{NMGM}(l, x_{\text{init}}, Q_lF(x)) - x\|^2 \\ &\quad\le\ \varphi((\sqrt{1-\underline{c}} + c_R)\,\|x_{\text{init}} - x\|\,, (1+c_R)\,\|x_{\text{init}} - x\|) \\ &\quad\le\ (1 - \underline{c} + \varepsilon^{2m}\underline{c} + \overline{c}\,c_R)\,\|x_{\text{init}} - x\|^2 \end{aligned} \tag{5.37}$$

for some $\overline{c} > 0$ independent of l, ε, and c_R provided that

$$4c_R^2(1+c_R)^2\varepsilon^{2m}(1+\varepsilon^m)^2 \le (1-c_R)^2(1-\varepsilon^{2m})^2(\underline{c} + 2c_R(1-\sqrt{1-\underline{c}}))\,.$$

This can always be achieved if c_R is sufficiently small which we will assume in the following.

Now, by our assumptions

$$1 - \underline{c} < c(m) := (1 - \tfrac{1}{m})(\underline{c}m)^{-\frac{1}{m-1}}\,, \tag{5.38}$$

which can be seen as follows: for

$$\psi(t) := 1 - t - (1 - \tfrac{1}{m})(mt)^{-\frac{1}{m-1}}$$

it holds that $\psi(\frac{1}{m}) = 0$ and that $\psi'(t) = -1 + (mt)^{-\frac{m}{m-1}} < 0$ for all $t > 1/m$. Thus, it follows with (5.19) that $\psi(\underline{c}) < 0$, which proves (5.38). Together with (5.23) we can conclude that

$$1 - \underline{c} + \varepsilon^{2m}\underline{c} + \overline{c}\,c_R = 1 - \underline{c} - c(m) + \overline{c}\,c_R + \varepsilon^2 \le \varepsilon^2$$

for c_R sufficiently small. Now the assertion follows with (5.37). □

We now consider the application of the nonlinear multigrid operator of Algorithm 5.3 in a nested iteration (cf. [59]) for finding an approximation $\tilde{x}_l^\delta$ to a solution x_l^δ of (5.1). For simplicity of notation, we consider only one iteration on each level:

Algorithm 5.5 (Nonlinear Full Multigrid Method, NFMGM)

(i) Set $\tilde{x}_1^\delta := \Phi_1(x_0, Q_1 y^\delta)$.

(ii) For $k = 2, \ldots, l$: Set $\tilde{x}_k^\delta :=$ NMGM$(k, \tilde{x}_{k-1}^\delta, Q_k y^\delta)$, where $\tilde{x}_{k-1}$ as in Algorithm 5.3 is chosen as $\tilde{x}_{k-1}^\delta$.

In the simple situation of V-cycle multigrid operators, this can be schematically represented as in Figure 5.2.

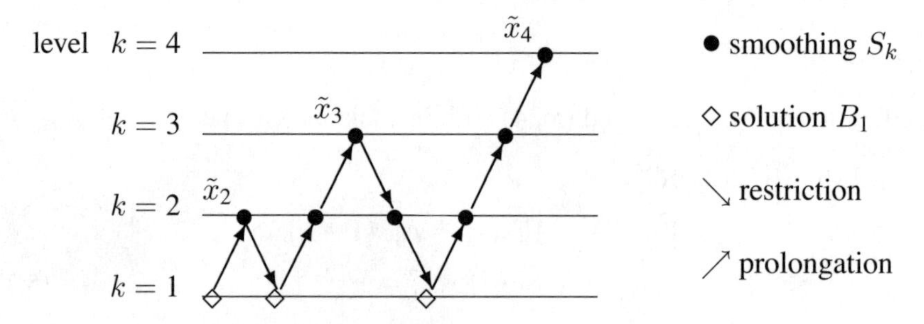

Figure 5.2 The FMGM with $m \equiv 1$

To prove convergence and convergence rates we assume a geometrically decreasing mesh size, i.e.,

$$h_l = h_0 \sigma^l \tag{5.39}$$

with some $h_0 > 0$ and $\sigma \in (0, 1)$.

As always, the iteration has to be stopped appropriately. We suggest to choose the stopping index $l_* = l_*(\delta)$ as in (5.9) or via

$$\eta \sigma^{l_* p} \tilde{\gamma}_{l_*}(\mu) \leq \delta < \eta \sigma^{lp} \tilde{\gamma}_l(\mu), \qquad 0 \leq l < l_*, \tag{5.40}$$

if the source condition (5.11) holds (compare (5.10)), where

$$\tilde{\gamma}_l(\mu) := \begin{cases} \varepsilon^l, & \varepsilon > \sigma^{2p\mu}, \\ l\sigma^{2\mu pl}, & \varepsilon = \sigma^{2p\mu}, \\ \sigma^{2\mu pl}, & \varepsilon < \sigma^{2p\mu}. \end{cases} \tag{5.41}$$

Theorem 5.6 *Let* (1.2), (5.39), *Assumption* 5.1, *and the assumptions of Proposition* 5.4 *hold with condition* (5.20) *replaced by*

$$16\frac{(1+c_R)^2}{1-c_R}\lambda_l \le 1\,. \tag{5.42}$$

Moreover, let $x^\dagger$ *be a solution of* (1.1) *satisfying* (5.6) *with* $\|x_0 - x^\dagger\|$ *sufficiently small.*

If l_* *is chosen according to* (5.9) *with* $\eta > 0$ *sufficiently small, then* $\tilde{x}_l^\delta$ *is well defined by Algorithm 5.5 and converges to* $x^\dagger$ *as* $\delta \to 0$.

If the source condition (5.11) *is satisfied for some* $\mu \in (0,1]$, *if* l_* *is chosen according to* (5.40), *and if* $\varepsilon < \sigma^{2p\mu}$, *then we obtain the convergence rate*

$$\|\tilde{x}_{l_*}^\delta - x^\dagger\| = O(\delta^{\frac{2\mu}{2\mu+1}})\,.$$

In the noise free case ($\delta = 0$), $\tilde{x}_l$ *converges to* $x^\dagger$ *as* $l \to \infty$ *and, if* (5.11) *holds with* $\mu \in (0,1]$, *the rate*

$$\|\tilde{x}_l - x^\dagger\| = O(\tilde{\gamma}_l(\mu))$$

is obtained.

Proof. First of all, we want to mention that we may assume that the solution x_l^δ of (5.1) not only exists in $\mathcal{B}_{2\rho}(x_0)$ but also in $\mathcal{B}_{\tilde{\rho}}(x_0)$ with possibly slightly adjusted constants $c_1^{rd}, c_2^{rd}, c_3^{rd}$ and level bound $\tilde{l}^{rd}$ in Proposition 5.2.

We will now show by induction that $\tilde{x}_l^\delta$ satisfies (5.20) for all $l \le \tilde{l}^{mg}$, where

$$\tilde{l}^{mg} := \max\{\tilde{l} \le \tilde{l}^{rd} \,:\, c_1^{mg}\|x_0 - x^\dagger\| + c_2^{mg}\delta\sigma^{-lp} \le \tfrac{1}{2}\tilde{\rho} \text{ for all } l \le \tilde{l}\} \tag{5.43}$$

with

$$c_1^{mg} := \frac{\max\{1, (1+\varepsilon)c_1^{rd}\}}{1-\varepsilon} \quad \text{and} \quad c_2^{mg} := \frac{(1+\varepsilon)c_1^{rd}\tilde{c}_2}{h_0^p(1-\varepsilon\sigma^p)}\,, \tag{5.44}$$

$\tilde{\rho}$ is as in (5.21), h_0 and σ are as in (5.39), $\tilde{c}_2$ is as in Assumption 5.1 (ii), and ε is as in (5.23). Note that, due to (5.19), $\varepsilon < 1$. Thus, Proposition 5.4 is applicable then.

We only show the general step: using Algorithm 5.5, (5.1), and Proposition 5.4, we obtain the estimate

$$
\begin{aligned}
\|\tilde{x}_l^\delta - x^\dagger\| &\leq \|\tilde{x}_l^\delta - x_l^\delta\| + \|x_l^\delta - x^\dagger\| \\
&\leq \varepsilon(\|\tilde{x}_{l-1}^\delta - x_{l-1}^\delta\| + \|x_{l-1}^\delta - x_l^\delta\|) + \|x_l^\delta - x^\dagger\| \\
&\leq \varepsilon^l \|x_0 - x_1^\delta\| + \sum_{k=1}^{l-1} \varepsilon^k \|x_{l-k}^\delta - x_{l-k+1}^\delta\| + \|x_l^\delta - x^\dagger\| \\
&\leq \varepsilon^l(\|P_1(x_0 - x^\dagger)\| + \|P_1(x_1^\delta - x^\dagger)\|) \\
&\quad + \sum_{k=1}^{l-1} \varepsilon^k (\|x_{l-k}^\delta - x^\dagger\| + \|x_{l-k+1}^\delta - x^\dagger\|) + \|x_l^\delta - x^\dagger\| \\
&\leq \varepsilon^l \|P_1(x_0 - x^\dagger)\| + (1+\varepsilon) \sum_{k=0}^{l-1} \varepsilon^k \|x_{l-k}^\delta - x^\dagger\|
\end{aligned}
$$

and hence Proposition 5.2 (ii), Assumption 5.1 (ii), (5.39), (5.43), and (5.44) yield

$$
\begin{aligned}
\|\tilde{x}_l^\delta - x^\dagger\| &\leq \varepsilon^l \|P_1(x_0 - x^\dagger)\| \\
&\quad + c_1^{mg}(1-\varepsilon) \sum_{k=0}^{l-1} \varepsilon^k \|(I - P_{l-k})(x_0 - x^\dagger)\| \qquad (5.45) \\
&\quad + c_2^{mg} \delta \sigma^{-lp} \\
&\leq c_1^{mg} \|x_0 - x^\dagger\| + c_2^{mg} \delta \sigma^{-lp} \leq \tfrac{1}{2}\tilde{\rho}. \qquad (5.46)
\end{aligned}
$$

Since $\|x_0 - x^\dagger\|$ is sufficiently small, we may assume that $\|x_0 - x^\dagger\| \leq \tilde{\rho}/4$, which together with (5.46) implies that $\|\tilde{x}_l^\delta - x_0\| \leq 3\tilde{\rho}/4$ and further with (5.42) that (5.20) is satisfied which makes Proposition 5.4 applicable.

Note that, due to (5.8), (5.39), and (5.43), l_* as chosen by (5.9) or (5.40) satisfies $l_* \leq \tilde{l}^{mg}$ if $\|x_0 - x^\dagger\|$ and $\eta > 0$ are sufficiently small, which we will assume in the following. This together with (5.45) and the fact that, due to (5.6), $\|(I - P_l)(x_0 - x^\dagger)\| \to 0$ as $l \to \infty$ already yields convergence of $\tilde{x}_{l_*}^\delta$ and $\tilde{x}_l$ towards $x^\dagger$.

Let us now assume that (5.11) holds for some $\mu \in (0, 1]$. Then we obtain with (4.119), Assumption 5.1 (i), (5.41), and (5.45) that

$$
\|\tilde{x}_l^\delta - x^\dagger\| = O(\tilde{\gamma}_l(\mu) + \delta\sigma^{-lp}),
$$

which proves the rate in the noise free case. In the noisy case, (5.40) implies that

$$
\|\tilde{x}_{l_*}^\delta - x^\dagger\| = O(\tilde{\gamma}_{l_*}(\mu)),
$$

since

$$\frac{\tilde{\gamma}_{l_*-1}(\mu)}{\tilde{\gamma}_{l_*}(\mu)} \leq c$$

for some positive constant c. If $\varepsilon < \sigma^{2p\mu}$ this finally implies the rate $O(\delta^{\frac{2\mu}{2\mu+1}})$. $\square$

Under the tangential cone condition (5.3) optimal convergence rates can also be achieved by means of the discrepancy principle (2.2) (see [93, Theorem 3]).

6 Level set methods

Level set methods have been applied successfully for *shape design, shape recovery*, and *free boundary value problems* (see, e.g., Burger and Osher [25] for a survey on level set methods for inverse problems applications).

Typically level set methods for the solution of inverse problems are based on evolutionary equations that are solved *via* a *time marching scheme*. The solution of this scheme at an appropriate time is considered an approximation of the inverse problem. For a-priori given time stepping the marching scheme can be considered as an iterative method.

In this chapter we derive in a systematic way iterative level set methods, iterative regularization techniques, and evolutionary equations for level set problems. All three concepts are formally linked, but it is notable that the analysis of the three concepts is significantly different.

In *shape recovery*, one is faced with the problem of finding the shape ∂D of an object D satisfying

$$F(\chi(D)) = y\,, \tag{6.1}$$

where $\chi(D)$ denotes the characteristic function of a set D.

Shape design consists in minimization of an energy functional $f(D)$ over a suitable class of objects D. Previously, shape recovery in inverse problems has been implemented by recovering parametrized shapes. There is rich literature on this subject: some relevant work on inverse problems applications can be found in [71, 72, 73, 74, 75, 78, 81, 95, 103]. The approach is limited to applications, where the number of connected components of D is known a-priori. *Topological optimization* is used for shape design where the number of connected components of the optimal shape is not specified a-priori (see, e.g., [9]).

Level set methods are capable of handling splitting and merging topologies and are currently a strong research topic for the numerical solution of inverse problems without a-priori specified numbers of connected components.

The basic idea of level set methods is to represent ∂D as the zero level set of a function ϕ, i.e., we have

$$\partial D = \{\phi = 0\} \quad \text{and} \quad \overline{D} = \{\phi \geq 0\}\,. \tag{6.2}$$

For some background on level set methods we refer, e.g., to Osher, Sethian, and Fedkiw [126, 127, 148].

We recall that for any measurable set $D \neq \emptyset$ satisfying

$$\overline{\mathrm{int}(D)} = \overline{D}\,,$$

∂D can be represented as the zero level set of a $W^{1,\infty}$ function (see, e.g., Delfour and Zolesio [35]). A level set function with $\partial D = \{\phi = 0\}$ is given, for instance, by the signed distance function

$$\phi = -d_{\overline{D}} + d_{\overline{CD}},$$

where $d_{\overline{D}}$ and $d_{\overline{CD}}$ denote the distance functions to $\overline{D}$ and $\overline{CD}$, respectively. Here, CD denotes the complementary set to D. The functions $d_{\overline{D}}$ and $d_{\overline{CD}}$ are uniformly Lipschitz continuous and satisfy $|\nabla d_{\overline{D}}| \leq 1$ and $|\nabla d_{\overline{CD}}| \leq 1$ (see [35]).

Let $\Omega \subset \mathbb{R}^n$ be a bounded domain with Lipschitz boundary, in which the object to be recovered is contained, then from the above considerations it follows that

$$d_{\overline{D}},\, d_{\overline{CD}} \in W^{1,\infty}(\Omega) \subset W^{1,2}(\Omega).$$

We consider level set functions in the Hilbert space $W^{1,2}(\Omega)$. In particular, all signed distance functions are contained in this space.

We define the discontinuous projection operator

$$P : W^{1,2}(\Omega) \to L^\infty(\Omega), \qquad \phi \mapsto \chi(\{\phi \geq 0\}).$$

In level set methods for shape recovery the problem of solving (6.1) is replaced by the problem of solving

$$F(P(\phi)) = y. \tag{6.3}$$

In the sequel we assume that the operator F is continuous from $L^1(\Omega)$ into the Hilbert space $L^2(\tilde{\Omega})$. The operator $F(P(\cdot))$ is difficult to deal with analytically, since it is the decomposition of a continuous operator and an operator that is not weakly closed on the set of functions with binary range $\{0, 1\}$ with respect to the L^2-topology. Note for instance that the characteristic functions $\rho_n := \sum_{i=1}^{2^n} \chi_{((2i-1)/2^n, 2i/2^n)}$ of the set $\bigcup_{i=1}^{2^n}((2i-1)/2^n, 2i/2^n)$ are weakly convergent to $1/2$, which is not a function with binary range $\{0, 1\}$.

For noisy data y^δ, we use the energy functional

$$J(\phi) := \tfrac{1}{2}\|F(P(\phi)) - y^\delta\|^2_{L^2(\tilde{\Omega})}. \tag{6.4}$$

In general, the functional J is not Fréchet-differentiable, but the subdifferential of J can be defined, at least in a formal setting. The concept of a subdifferential is outlined below. Formally, it reads as

$$\partial J(\phi) = P'(\phi)^* F'(P(\phi))^* (F(P(\phi)) - y^\delta), \tag{6.5}$$

where $F'(z)^*$ is the formal adjoint with respect to the spaces $L^2(\Omega)$ and $L^2(\tilde{\Omega})$, i.e.,

$$\langle F'(z)h, g \rangle_{L^2(\tilde{\Omega})} = \langle h, F'(z)^* g \rangle_{L^2(\Omega)}$$

for all $h \in L^2(\Omega)$ and $g \in L^2(\tilde{\Omega})$. Moreover, for $\phi, \tilde{\phi}, \hat{\phi} \in W^{1,2}(\Omega)$,

$$\langle P'(\phi)\hat{\phi}, \tilde{\phi} \rangle_{L^2(\Omega)} = \langle \hat{\phi}, P'(\phi)^* \tilde{\phi} \rangle_{W^{1,2}(\Omega)} .$$

The functional $F'(P(\phi(t)))P'(\phi(t))$ can be considered as the shape derivative of the functional F with respect to the interface given by the support of $P'(\phi)$. The right hand side of (6.5) is of a complex nature, since the functional derivative P' of P only exists in a distributional setting.

The Landweber iteration for the solution of (6.3) is given by

$$\phi_{k+1} \in \phi_k - \partial J(\phi_k) . \tag{6.6}$$

We use the symbol $\in$, since, in general, $\partial J(\phi_k)$ is set-valued. Therefore, for a practical realization of (6.6) a single element of $\partial J(\phi_k)$ has to be selected.

In the level set community, it is more convenient to base the consideration on a continuous regularization method instead of (6.6), namely

$$\frac{\partial \phi}{\partial t}(t) \in -\partial J(\phi)(t) , \quad t > 0 , \quad \text{and} \quad \phi(0) = \phi_0 , \tag{6.7}$$

and consider (6.6) as a method for numerically realizing (6.7).

Currently, for inverse problems applications, there is neither a convergence or stability analysis of (6.6) available nor results for existence and uniqueness of a solution of (6.7). However, as we will show below, the implicit time steps of (6.7) are well defined. It is worth noting that implicit time stepping methods lead to Newton type level set regularization models as they have been introduced by Burger [20].

In the remainder of this chapter we proceed as follows: we give an overview on existence results of time-dependent processes based on the concept of maximal monotone operators. By this argumentation we can highlight the difficulties in performing a rigorous analysis of equations of type (6.7). Moreover, the theory of maximal monotone operators and evolutionary processes reveals some synergies with the analysis of the Landweber iteration. This is expected, since the Landweber iteration can be considered as an explicit time stepping algorithm for the solution of the according parabolic process.

6.1 Continuous regularization

Continuous regularization for the solution of the nonlinear operator equation (1.1) consists in solving the time-dependent equation

$$\frac{\partial x}{\partial t}(t) = F'(x(t))^*(y^\delta - F(x(t))) , \quad t > 0 , \quad \text{and} \quad x(0) = x_0 , \tag{6.8}$$

up to a certain time T. For inverse and ill-posed problems, under certain assumptions, $x(T)$ is a stable approximation of the solution of (1.1). The stopping time T plays the role of a regularization parameter.

The *explicit Euler method* for approximating (6.8) at time $t + \Delta t$ is given by

$$x(t + \Delta t) \approx x(t) + \Delta t F'(x(t))^*(y^\delta - F(x(t))) \,, \tag{6.9}$$

which can be considered as an iteration step of a Landweber iteration, where Δt plays the role of ω (see (2.24); cf. [10]).

The *implicit Euler method* for solving (6.8) consists in solving the Euler equation

$$(x - x(t)) - \Delta t F'(x)^*(y^\delta - F(x)) = 0 \tag{6.10}$$

and denoting a solution by $x(t + \Delta t)$. The solution of (6.10) approximates the solution of (6.8) at $t + \Delta t$. We recall that, if the functional $\|F(x) - y^\delta\|^2$ is convex, then the solution of (6.10) is the unique minimizer of the functional

$$\|x - x(t)\|^2 + \Delta t \|F(x) - y^\delta\|^2. \tag{6.11}$$

This means that the implict Euler method corresponds to iterative Tikhonov regularization, where $1/\Delta t$ plays the role of α.

For an efficient minimization of (6.11) the Gauss–Newton method can be used (compare (4.36)), which consists in putting $x^{(0)} := x(t)$ and

$$\begin{aligned} x^{(k+1)} \;=\; & x^{(k)} + \Big(I + \Delta t F'(x^{(k)})^* F'(x^{(k)})\Big)^{-1} \\ & \Big(\Delta t F'(x^{(k)})^*(y^\delta - F(x^{(k)})) + (x^{(0)} - x^{(k)})\Big). \end{aligned} \tag{6.12}$$

Assume that *one* iteration of (6.12) is sufficient to calculate a close approximation of $x(t + \Delta t)$, then

$$\begin{aligned} x(t + \Delta t) \;\approx\; & x(t) + \Delta t \Big(I + \Delta t F'(x(t))^* F'(x(t))\Big)^{-1} \\ & F'(x(t))^*(y^\delta - F(x(t))) \end{aligned} \tag{6.13}$$

(see also [92]). Comparing this time-marching scheme with the explicit Euler method (6.9), it is quite paradox to see that both methods can be considered as approximations of $x(t+\Delta t)$, although the right hand sides are formally completely different. It is of course no contradiction, since our derivation of (6.13) is based on the assumption that one step of (6.12) provides a reasonable approximation of the minimizer of the functional (6.11). We also stress the fact that, by simply setting $x_{k+1}^\delta := x(t + \Delta t)$ and $x_k^\delta := x(t)$, (6.13) looks like the Levenberg–Marquardt method (cf. (4.2)).

The above considerations indicate that both, Landweber iteration and the Levenberg–Marquardt scheme, can be considered as *time marching schemes*, in which case they are considered as approximations of the solution of the continuous problem (6.8). Note that the existence of a solution of (6.8) is not required in the analysis of iterative regularization methods, but is needed in order to show that time marching schemes approximate the solution of (6.8).

A classical concept for solutions of (6.8) is provided by *semigroup* theory based on the theory of *maximal monotonicity* and *subgradients*. These concepts are outlined below. For a detailed discussion on semigroup theory we refer to Brézis [15], Zeidler [161], and Evans [48].

Let $\mathcal{X}$ be a real Banach space. We recall that a functional $J : \mathcal{X} \to \mathbb{R} \cup \{+\infty\}$ is *convex* if

$$J(tx + (1-t)\tilde{x}) \leq tJ(x) + (1-t)J(\tilde{x}) \quad \text{for all} \quad x, \tilde{x} \in \mathcal{X} \quad \text{and} \quad t \in [0,1]\,.$$

It is *proper* if $J \not\equiv +\infty$. In particular, it follows from the definition of a convex functional that

$$\mathcal{D}(J) = \{x \in \mathcal{X} \,:\, J(x) < +\infty\}$$

is convex.

A functional J defined on $\mathcal{X}$ is called *lower semicontinuous* if $x_k \to x$ implies that

$$J(x) \leq \liminf_{k\to\infty} J(x_k)\,.$$

The *subdifferential* ∂J of a convex, proper functional J defined on $\mathcal{X}$ is defined for $x \in \mathcal{X}$ satisfying $J(x) \neq +\infty$ as follows: $\partial J(x)$ consists of all elements $\rho \in \mathcal{X}^*$ (the dual space of $\mathcal{X}$) satisfying

$$J(\tilde{x}) \geq J(x) + \rho(\tilde{x} - x) \qquad \text{for all} \quad \tilde{x} \in \mathcal{X}\,. \tag{6.14}$$

In particular, if $\mathcal{X}$ is a real Hilbert space, then $\rho \in \partial J(x) \subset \mathcal{X}$ if $J(x) \neq +\infty$ and

$$J(\tilde{x}) \geq J(x) + \langle \rho, \tilde{x} - x \rangle \qquad \text{for all} \quad \tilde{x} \in \mathcal{X}\,. \tag{6.15}$$

We set

$$\mathcal{D}(\partial J) = \{x \in \mathcal{X} \,:\, J(x) \neq +\infty \text{ and } \partial J(x) \neq \emptyset\}\,.$$

Theorem 6.1 *Assume that $\mathcal{X}$ is a real Hilbert space and that $J : \mathcal{X} \to \mathbb{R} \cup \{+\infty\}$ is convex, proper and lower semicontinuous. Then it holds:*

(i) $\mathcal{D}(\partial J) \subset \mathcal{D}(J)$.

(ii) Monotonicity: *If $\rho \in \partial J(x)$ and $\tilde{\rho} \in \partial J(\tilde{x})$, then $\langle \rho - \tilde{\rho}, x - \tilde{x} \rangle \geq 0$.*

(iii) $J(x^*) = \min_{x \in \mathcal{X}} J(x)$ *if and only if* $0 \in \partial J(x^*)$.

(iv) For each $x_0 \in \mathcal{X}$ *and* $\lambda > 0$*, the problem*

$$x + \lambda \partial J(x) \ni x_0 \tag{6.16}$$

has a unique solution $u \in \mathcal{D}(\partial J)$. (6.16) *is an* inclusion equation *and the left hand side is, in general, a set consisting of* $x + \lambda\rho = x_0$ *with* $x \in \mathcal{D}(\partial J)$ *and* $\rho \in \partial J(x)$.

Proof. See [48, p. 524 ff]. □

Example 6.2 Below we give two examples for subgradients:

(i) Let C be a nonempty, closed, convex subset of a real Hilbert space $\mathcal{X}$. We define $\chi(x) = 0$ if $x \in C$ and $\chi(x) = +\infty$ if $x \notin C$. Obviously, the subgradient of χ is only defined for $x \in C$ and according to (6.15) an element ρ of the subgradient has to satisfy

$$\chi(\tilde{x}) \geq \langle \rho, \tilde{x} - x \rangle \qquad \text{for all} \quad \tilde{x} \in \mathcal{X}$$

or equivalently

$$0 \geq \langle \rho, \tilde{x} - x \rangle \qquad \text{for all} \quad \tilde{x} \in C .$$

In particular, it follows from this inequality that $\rho = 0 \in \partial\chi(x)$ for every $x \in C$. If $x \in \text{int}(C)$, then there exists an $\varepsilon > 0$ such that

$$0 \geq \langle \rho, \tilde{x} - x \rangle \qquad \text{for all} \quad \tilde{x} \in \mathcal{X} \;\text{ with }\; \|\tilde{x} - x\| < \varepsilon .$$

This shows that $\{0\} = \partial\chi(x)$ for $x \in \text{int}(C)$.

(ii) Let $J(x) := \|F(x) - y^\delta\|^2/2$ be convex on a real Hilbert space $\mathcal{X}$, with F Fréchet-differentiable. Then the subgradient $\partial J(x)$ is single-valued and is given by $F'(x)^*(F(x) - y^\delta)$.

Let $J : \mathcal{X} \to \mathbb{R} \cup \{+\infty\}$ be a convex, proper, and lower semicontinuous functional on a real Hilbert space $\mathcal{X}$. For each $\lambda > 0$ we define the *nonlinear resolvent* $J_\lambda : \mathcal{X} \to \mathcal{D}(\partial J)$ by setting

$$J_\lambda(x_0) = x ,$$

where x is the unique solution of

$$x + \lambda\partial J(x) \ni x_0 ,$$

which exists due to Theorem 6.1 (iv). Note that $J_\lambda(x_0)$ can as well be characterized as the minimizer of the functional

$$x \to \tfrac{1}{2}\|x - x_0\|^2 + \lambda J(x)\,.$$

Thus, the nonlinear resolvent corresponds to a Tikhonov type regularized solution.

For each $\lambda > 0$ the *Yosida approximation* $A_\lambda : \mathcal{X} \to \mathcal{X}$ is defined by

$$A_\lambda(x_0) := \frac{x_0 - J_\lambda(x_0)}{\lambda}\,.$$

We denote by $2^{\mathcal{X}}$ the power set of a real Hilbert space $\mathcal{X}$. For a set-valued operator $A : \mathcal{X} \to 2^{\mathcal{X}}$ we denote by

$$\mathcal{D}(A) = \{x \in \mathcal{X} : A(x) \neq \emptyset\}$$

its *domain* and by

$$\mathcal{R}(A) = \bigcup_{x \in \mathcal{X}} A(x)$$

its *range*. The operator A is called *single-valued* if $A(x)$ consists of a single element for all $x \in \mathcal{D}(A)$. The operator A is called *monotone* if

$$\langle z_x - z_{\tilde{x}}, x - \tilde{x} \rangle \geq 0 \qquad \text{for all} \quad \tilde{x} \in \mathcal{D}(A)\,,\ z_{\tilde{x}} \in A(\tilde{x})\,, \tag{6.17}$$

and for all $x \in \mathcal{D}(A)$, $z_x \in A(x)$. A monotone operator A is called *maximal monotone* if for every pair $(x, z_x) \in \mathcal{X} \times \mathcal{X}$ which satisfies (6.17) for every $\tilde{x} \in \mathcal{D}(A)$ and $z_{\tilde{x}} \in A(\tilde{x})$ it holds that $x \in \mathcal{D}(A)$ and $z_x \in A(x)$. The terminology maximal monotone operator is motivated from the fact that it is a monotone operator that has no proper extension. An extension $\overline{A}$ of A satisfies that the graph of A is a real subset of the graph of $\overline{A}$. In particular this means that $\mathcal{D}(A) \subset \mathcal{D}(\overline{A})$. For instance, the function in the center of Figure 6.1 is monotone but not maximal monotone, since it has a proper extension, which is plotted in the right of Figure 6.1.

A continuous and monotonically increasing function $f : \mathbb{R} \to \mathbb{R}$ is a typical example for a maximal monotone operator. For some other examples of maximal monotone operators see Example 6.7 below.

The following theorem holds for subdifferentials:

Theorem 6.3 *Assume that $\mathcal{X}$ is a real Hilbert space and that $J : \mathcal{X} \to \mathbb{R} \cup \{+\infty\}$ is convex, proper, and lower semi-continuous. Then ∂J is maximal monotone.*

Proof. See, e.g., Zeidler [161, p. 845]. □

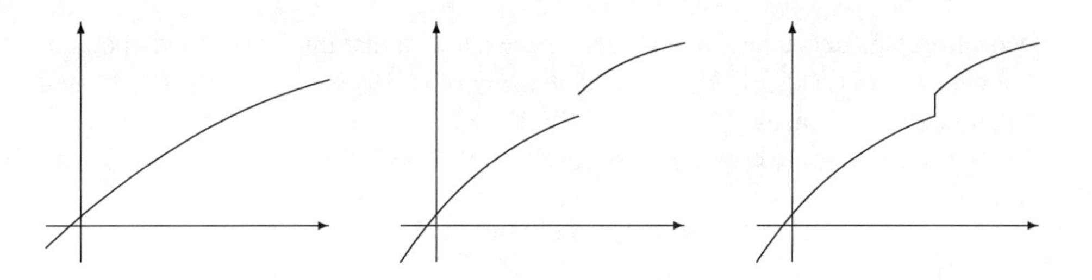

Figure 6.1 *left:* a maximal monotone function; *center* and *right:* a function which is not maximal monotone and its maximal monotone extension

Our intention is to cite an existence result for the ordinary differential equation

$$\frac{\partial x}{\partial t}(t) \in -\partial J(x(t)), \quad t > 0, \quad \text{and} \quad x(0) = x_0 . \tag{6.18}$$

Theorem 6.4 *Let $\mathcal{X}$ be a real Hilbert space and $J : \mathcal{X} \to \mathbb{R} \cup \{+\infty\}$ be convex, proper and lower semicontinuous. Moreover, assume that ∂J is densely defined, i.e., $\overline{\mathcal{D}(\partial J)} = \mathcal{X}$. Then for each $x_0 \in \mathcal{D}(\partial J)$ there exists a unique function*

$$x \in C([0,\infty); \mathcal{X}) \qquad \textit{with} \qquad \frac{\partial x}{\partial l} \in L^\infty((0,\infty); \mathcal{X})$$

satisfying

$$\begin{aligned} x(0) &= x_0 , \\ x(t) &\in \mathcal{D}(\partial J) \quad \textit{for each} \quad t > 0 , \\ \frac{\partial x}{\partial t} &\in -\partial J(x(t)) \quad \textit{for almost every} \quad t > 0 . \end{aligned}$$

(For the definition of the spaces $C([0,\infty); \mathcal{X})$ and $L^\infty((0,\infty); \mathcal{X})$ see [48].)

Proof. See Evans [48, p. 529 ff]. □

This result can be generalized from gradient flow equations to equations with maximal monotone operators.

For the analysis of evolutionary processes with maximal monotone operators

$$\frac{\partial x}{\partial t}(t) \in -A(x(t)) + f(t), \quad 0 < t \leq T, \quad \text{and} \quad x(0) = x_0 , \tag{6.19}$$

typically the following notion of a solution is used (see Brezis [15, p. 65]):

Definition 6.5 Let $\mathcal{X}$ be a real Hilbert space and A a maximal monotone operator. A function $x \in C([0,T];\mathcal{X})$ is called *solution* of (6.19) if $f \in L^1((0,T);\mathcal{X})$ and if there exist sequences $f_n \in L^1((0,T);\mathcal{X})$ and $x_n \in C([0,T];\mathcal{X})$, where x_n is absolutely continuous in $(0,T)$ and $x_n(t) \in \mathcal{D}(A)$ for all $t \in (0,T)$, satisfying:

$$\begin{aligned} &f_n \to f \quad \text{with respect to the } L^1((0,T);\mathcal{X})\text{-norm}\,, \\ &\frac{\partial x_n}{\partial t} + A(x_n) \ni f_n\,, \quad t > 0\,, \\ &\|x_n(t) - x(t)\| \to 0 \quad \text{uniformly in} \quad [0,T]\,. \end{aligned}$$

Theorem 6.6 *Assume that A is maximal monotone on $\mathcal{X}$. If $f \in L^1((0,T);\mathcal{X})$ and $x_0 \in \overline{\mathcal{D}(A)}$, then there exists a unique solution of* (6.19).

Proof. See Brezis [15, Theorem 3.4]. □

Example 6.7 We apply the theorems above to some examples:

(i) Let C and χ be as in Example 6.2 (i). Then the subdifferential $\partial\chi$ is maximal monotone. Therefore, it follows with Theorem 6.6 that for every $x_0 \in \overline{C}$ the differential equation

$$\frac{\partial x}{\partial t} \in -\partial\chi(x)\,, \quad t > 0\,, \quad \text{and} \quad x(0) = x_0\,,$$

has a unique solution.

(ii) Let $K : \mathcal{X} \to \mathcal{Y}$ be a bounded linear operator between real Hilbert spaces $\mathcal{X}$ and $\mathcal{Y}$. Then the functional $J(x) := \|Kx - y^\delta\|^2/2$ is convex, proper, and lower semi-continuous and the subgradient $\partial J(x)$ is given by $K^*(Kx - y^\delta)$. According to Theorem 6.3, $\partial J(x)$ is maximal monotone. Due to Theorem 6.4, the problem

$$\frac{\partial x}{\partial t}(t) = -K^*(Kx - y^\delta)\,, \quad t > 0\,, \quad \text{and} \quad x(0) = x_0\,, \tag{6.20}$$

has a unique solution $x^\delta \in C([0,\infty);\mathcal{X})$ with $\frac{\partial x^\delta}{\partial t} \in L^\infty((0,\infty);\mathcal{X})$ for any $x_0 \in \mathcal{X}$. Thus, for linear inverse problems continuous regularization attains a solution.

(iii) Let $F : \mathcal{D}(F) \subset \mathcal{X} \to \mathcal{Y}$ be a nonlinear continuously Fréchet-differentiable operator between two real Hilbert spaces $\mathcal{X}$ and $\mathcal{Y}$, which satisfies (2.3) and

$$\begin{aligned} &\|F(x) - F(\tilde{x}) - F'(x)(x - \tilde{x})\| \le c\|F(x) - F(\tilde{x})\|^2, \\ &x, \tilde{x} \in \mathcal{B}_{2\rho}(x_0) \subseteq \mathcal{D}(F)\,, \end{aligned} \tag{6.21}$$

for some $c > 0$. Note, that in contrast to (2.4) we assume now an exponent 2 in the right hand side of the inequality. Hence, this condition is even stronger than (2.25).

We assume that there exists a solution of $F(x) = y$ in $\mathcal{B}_\rho(x_0)$ which is denoted by x_* and define the operator

$$G(x) := \begin{cases} F'(x)^*(F(x) - y^\delta), & x \in \mathcal{B}_{2\rho}(x_0), \\ \emptyset, & x \notin \mathcal{B}_{2\rho}(x_0), \end{cases}$$

with $\|y^\delta - y\| \le \delta$ (see (1.2)). Since it follows from (2.3) that

$$\begin{aligned} \|F(x) - y\| &\le \int_0^1 \|F'(x_* + t(x - x_*))\| \, \|x - x_*\| \, dt \\ &\le \|x - x_*\| \le \|x - x_0\| + \|x_0 - x_*\| \le 3\rho \end{aligned}$$

for all $x \in \mathcal{B}_{2\rho}(x_0)$, we obtain for all $x, \tilde{x} \in \mathcal{B}_{2\rho}(x_0)$ that

$$\begin{aligned} &\langle G(x) - G(\tilde{x}), x - \tilde{x} \rangle \\ &\quad = \langle F(x) - y^\delta, F'(x)(x - \tilde{x}) \rangle - \langle F(\tilde{x}) - y^\delta, F'(\tilde{x})(x - \tilde{x}) \rangle \\ &\quad = \langle F(x) - y^\delta, F(x) - F(\tilde{x}) \rangle - \langle F(\tilde{x}) - y^\delta, F(x) - F(\tilde{x}) \rangle \\ &\qquad - \langle F(x) - y^\delta, F(x) - F(\tilde{x}) - F'(x)(x - \tilde{x}) \rangle \\ &\qquad + \langle F(\tilde{x}) - y^\delta, F(x) - F(\tilde{x}) - F'(\tilde{x})(x - \tilde{x}) \rangle \\ &\quad \ge (1 - 2c(3\rho + \delta)) \|F(x) - F(\tilde{x})\|^2 \ge 0 \end{aligned}$$

if $2c(3\rho + \delta) \le 1$. Thus, under this assumption, G is monotone on $\mathcal{B}_{2\rho}(x_0)$. With G we identify the set-valued mapping that maps x onto $\emptyset$ if $x \notin \mathcal{B}_{2\rho}(x_0)$. This mapping is monotone on $\mathcal{X}$. According to [161, Theorem 32.M], there exists a maximal monotone extension $\overline{G} : \mathcal{X} \to 2^{\mathcal{X}}$ for which $\mathcal{D}(G) = \mathcal{D}(\overline{G}) = \mathcal{B}_{2\rho}(x_0)$ and the graph of G is a subset of the graph of $\overline{G}$.

The proof of the maximal monotone extension requires Zorn's Lemma and is, thus, not constructive. Here, since G is single-valued, we can specify $\overline{G}$ in $\operatorname{int}(\mathcal{B}_{2\rho}(x_0))$: for $x \in \operatorname{int}(\mathcal{B}_{2\rho}(x_0))$ let $\tilde{x} = x + th$, where $\|h\| \le 2\rho - \|x - x_0\|$ and $t \in (-1, 1)$. Then it is evident that $\tilde{x} \in \operatorname{int}(\mathcal{B}_{2\rho}(x_0))$. Since $\overline{G}$ is monotone, we have by definition that

$$\langle \rho_x - \rho_{\tilde{x}}, x - \tilde{x} \rangle \ge 0 \quad \text{for all} \quad \rho_x \in \overline{G}(x) \text{ and } \rho_{\tilde{x}} \in \overline{G}(\tilde{x}).$$

Since $\overline{G}$ is a maximal monotone extension, we have $G(x) \subset \overline{G}(x)$ and, therefore, that

$$\langle \rho_x - G(\tilde{x}), x - \tilde{x} \rangle = -t \langle \rho_x - G(x + th), h \rangle \ge 0 \quad \text{for all} \quad \rho_x \in \overline{G}(x).$$

Since the above inequality holds for all $t \in (-1, 1)$, we obtain together with the continuity of G in $\mathrm{int}(\mathcal{B}_{2\rho}(x_0))$, which follows from the continuity of F', that $\langle \rho_x - G(x), h \rangle = 0$. Since this equality holds for all h sufficiently small, we have $\rho_x = G(x)$.

Together with Theorem 6.6 we obtain that the differential equation

$$\frac{\partial x}{\partial t} \in -\overline{G}(x)\,, \quad 0 < t \leq T\,, \quad \text{and} \quad x(0) = x_0\,,$$

has a unique solution. This shows that there exists a solution for continuous regularization if (6.21) and (2.3) hold. We emphasize that the existence of a solution of continuous regularization has not been proven under the weaker convergence condition (2.4) for Landweber iteration.

For continuous regularization applied to linear problems (see Example 6.7 (ii)), Tautenhahn [150] provided the following results, which can be derived from general results of Vainikko and Veretennikov [153]:

(i) In the case of noise free data, i.e., $y^\delta = y$, the solution $x(T)$ of (6.20) satisfies $x(T) \to x^\dagger$ for $T \to \infty$.

(ii) A-priori parameter selection criterion: let y^δ satisfy (1.2). Then for $T := T(\delta)$ satisfying $T(\delta) \to \infty$ and $\delta^2 T(\delta) \to 0$ as $\delta \to 0$, it holds that $x^\delta(T) \to x^\dagger$.

(iii) A-posteriori parameter strategy: let y^δ satisfy (1.2) and assume that

$$\|Kx^\dagger - y^\delta\| > \tau\delta$$

for some $\tau > 1$. Then for T_* satisfying the discrepancy principle

$$\|Kx^\delta(T_*) - y^\delta\| = \tau\delta$$

it holds that

$$x^\delta(T_*) \to x^\dagger\,.$$

The difficulty associated with the analysis of gradient flow equations for nonlinear inverse problems is the existence theory of a solution. If one assumes the nonlinearity condition (2.4), then maximal monotonicity of $G(x) := F'(x)^*(F(x) - y^\delta)$ is not guaranteed and, consequently, classical existence theory of parabolic processes (as outlined above) is not available. Only if (6.21) holds, existence of the solution of the evolutionary process is guaranteed. Tautenhahn [150] carried over the results for continuous regularization from linear inverse problems to nonlinear ones. However, there the theorectical results on convergence and stability are based on the assumption that a solution of the gradient flow equation exists. This is of course not a trivial assumption.

6.2 Level set regularization

Santosa [140] introduced level set methods based on the gradient descent flow equation (6.7) for the level set function. Several analytical difficulties show up for this approach. To name but a few:

(i) The energy functional $J(\phi)$ defined in (6.4) is in general not convex.

(ii) The definition of J involves the evaluation of a discontinuous operator P. To make the functional J formally well defined, it has to be relaxed. This is outlined below.

(iii) The operator $\phi \mapsto \partial J(\phi)$ is not maximal monotone and, therefore, nonlinear semigroup theory is not applicable. It neither satisfies (6.21) nor (2.4). Consequently, the theory of continuous regularization is not applicable either.

In the following we mimic nonlinear semigroup theory and consider implicit time steps of (6.7) consisting in the solution of

$$\frac{\phi(t+\Delta t) - \phi(t)}{\Delta t} \in -\partial J(\phi)(t+\Delta t)\,.$$

The solution of this equation may not be unique, and some elements could be critical points. We restrict our attention only to a solution $\phi(t+\Delta t)$ that is a global minimizer of

$$\mathcal{J}_\lambda(\phi) := \tfrac{1}{2}\lambda \|F(P(\phi)) - y^\delta\|^2_{L^2(\tilde{\Omega})} + \tfrac{1}{2}\|\phi - \phi_0\|^2_{W^{1,2}(\Omega)}\,, \tag{6.22}$$

with $\lambda = 1/\Delta t$ and $\phi_0 = \phi(t)$. As a prerequisite step we analyze the functionals $\mathcal{J}_\lambda$ analytically and prove existence of a minimizer, approximation properties and discuss the numerical realization. The analysis of the according time-dependent process (6.7) is still open.

Application of the Gauss–Newton method for minimizing (6.22) and assuming similar to (6.12) and (6.13) that one iteration step is sufficient to obtain a reasonable approximation of $\phi(t+\Delta t)$ we can again proceed iteratively and get the following approximation of (6.7):

$$\begin{aligned}\phi(t+\Delta t) &= \phi(t) + \Delta t(I + \Delta t B(t)^* B(t))^{-1} B(t)^*(y^\delta - F(\phi(t)))\\ &\text{with}\quad B(t) := F'(P(\phi(t)))P'(\phi(t))\,.\end{aligned} \tag{6.23}$$

As in (6.13), by a change of variable, $\phi_{k+1} = \phi(t+\Delta t)$ and $\phi_k = \phi(t)$, we recover the Levenberg–Marquardt method.

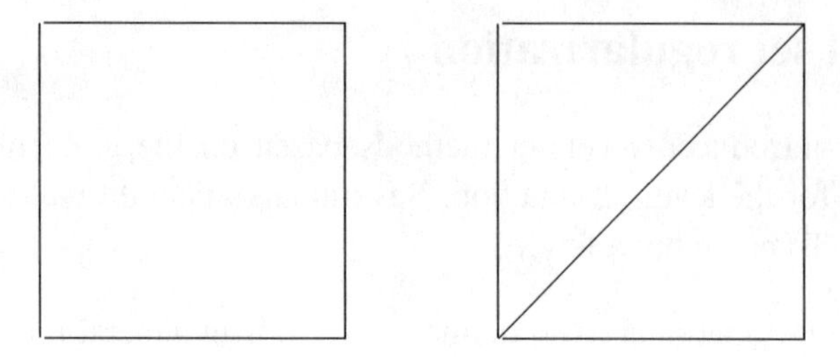

Figure 6.2 *left:* a domain with minimal length boundary; *right:* the domain on the right is equal to the domain on the left almost everywhere, but the boundary has different length

Two domains D_1 and D_2 are considered to be identical if the characteristic functions are identical almost everywhere. Of course, we want to exclude ambiguous cases and aim to recover only a domain with minimal boundary length.

In Figure 6.2, we have plotted two domains which are identical almost everywhere but the length of the boundary is different. To exclude this ambiguity the additional penalization term $|P(\phi)|_{\mathrm{BV}(\Omega)}$ is added to the functional $\mathcal{J}_\lambda$. Here, $\mathrm{BV}(\Omega)$ is the space of functions of bounded variation with its usual semi norm $|\cdot|_{\mathrm{BV}(\Omega)}$ (see [49]). This ensures the selection of a domain with minimal boundary length among all almost everywhere equivalent sets. The use of this penalization term and the above argumentation is inspired by a series of papers by Kohn and Strang [97, 98, 99], who discussed the necessity of using this penalization term for shape optimization and recovery. Thus, we consider the modified functional

$$\mathcal{J}_{\lambda,\beta}(\phi) = \tfrac{1}{2}\lambda \|F(P(\phi)) - y^\delta\|^2_{L^2(\tilde{\Omega})} + \beta\, |P(\phi)|_{\mathrm{BV}(\Omega)} + \tfrac{1}{2}\, \|\phi - \phi_0\|^2_{W^{1,2}(\Omega)}\,,$$

where β is a fixed positive parameter. In order to cope with difficulties in the analysis of this functional, due to the discontinuous operator P, we study the equivalent problem of minimizing

$$\mathcal{J}_{\lambda,\beta}(z,\phi) = \tfrac{1}{2}\lambda \|F(z) - y^\delta\|^2_{L^2(\tilde{\Omega})} + \beta\, |z|_{\mathrm{BV}(\Omega)} + \tfrac{1}{2}\, \|\phi - \phi_0\|^2_{W^{1,2}(\Omega)} \tag{6.24}$$

subject to the constraint $z = P(\phi)$.

Here the difficulties arising from the discontinuous operator become apparent. Since the operator P is discontinuous, the equation $z = P(\phi)$ has to be interpreted set-valued as outlined below. To this end, we consider continuous approximations of P. The operators

$$P_\varepsilon(\phi) := \begin{cases} 0\,, & \phi < -\varepsilon\,, \\ 1 + \varepsilon^{-1}\phi\,, & \phi \in [-\varepsilon, 0]\,, \\ 1\,, & \phi > 0\,, \end{cases} \tag{6.25}$$

satisfy

$$\|P_\varepsilon(\tilde{\phi}) - P_\varepsilon(\phi)\|_{L^1(\Omega)} \le \varepsilon^{-1}\,\mathrm{meas}(\Omega)^{\frac{1}{2}}\,\|\tilde{\phi} - \phi\|_{L^2(\Omega)}$$

and are, therefore, continuous from $L^2(\Omega)$ into $L^1(\Omega)$. Of course, any other choice of a family of approximations being continuous from $L^2(\Omega)$ into $L^1(\Omega)$ is suitable as well.

Next we define generalized minimizers and *minimal admissible pairs* of the functional $\mathcal{J}_{\lambda,\beta}$.

Definition 6.8 (i) A pair of functions $(z, \phi) \in L^1(\Omega) \times W^{1,2}(\Omega)$ is called *admissible* if there exists a sequence $\{\phi_k\}$ of functions in $W^{1,2}(\Omega)$ such that $\|\phi_k - \phi\|_{L^2(\Omega)} \to 0$ and if there exists a sequence $\{\varepsilon_k\}$ of positive numbers converging to zero such that $\|P_{\varepsilon_k}(\phi_k) - z\|_{L^1(\Omega)} \to 0$.

(ii) A *generalized solution* of (6.3) is an admissible pair $(z, \phi) \in L^1(\Omega) \times W^{1,2}(\Omega)$ satisfying $F(z) = y$.

(iii) A *generalized minimizer* of $\mathcal{J}_{\lambda,\beta}$ is an admissible pair of functions (z, ϕ) minimizing

$$\widetilde{\mathcal{J}}_{\lambda,\beta}(z,\phi) := \tfrac{1}{2}\lambda\|F(z) - y^\delta\|^2_{L^2(\tilde{\Omega})} + \rho(z,\phi) \tag{6.26}$$

over the set of admissible pairs, where

$$\rho(z,\phi) := \inf \liminf_{k\to\infty}\{\beta\,|P_{\varepsilon_k}(\phi_k)|_{\mathrm{BV}(\Omega)} + \tfrac{1}{2}\,\|\phi_k - \phi_0\|^2_{W^{1,2}(\Omega)}\} \tag{6.27}$$

is the *relaxation* of

$$\beta\,|P(\phi)|_{\mathrm{BV}(\Omega)} + \tfrac{1}{2}\,\|\phi - \phi_0\|^2_{W^{1,2}(\Omega)}$$

and the infimum is taken over all sequences $\{\varepsilon_k\}$ and $\{\phi_k\}$ as in (i).

(iv) A *minimal admissible pair* $(z^\dagger, \phi^\dagger)$ satisfies $F(z^\dagger) = y$ and

$$\rho(z^\dagger,\phi^\dagger) = \rho_{\min} := \inf\{\rho(z,\phi) \,:\, (z,\phi) \text{ is admissible and } F(z) = y\}\,. \tag{6.28}$$

Let (z, ϕ) be admissible and $\rho(z, \phi) < \infty$, then there exists a sequence $\{\phi_k\}$ of $W^{1,2}(\Omega)$-functions converging to ϕ in $L^2(\Omega)$ and there exists a sequence of positive numbers $\{\varepsilon_k\}$ converging to zero and satisfying $P_{\varepsilon_k}(\phi_k) \to z$ in $L^1(\Omega)$. From the weak lower semicontinuity of the $\mathrm{BV}(\Omega)$-seminorm it follows that

$$|z|_{\mathrm{BV}(\Omega)} \le \liminf_{k\to\infty} |P_{\varepsilon_k}(\phi_k)|_{\mathrm{BV}(\Omega)}\,.$$

In particular, this shows that under the assumption $\rho(z,\phi) < \infty$, we have $z \in \mathrm{BV}(\Omega)$. The above calculations also show that

$$\beta|z|_{\mathrm{BV}(\Omega)} + \tfrac{1}{2}\|\phi - \phi_0\|^2_{W^{1,2}(\Omega)} \le \rho(z,\phi)\,. \tag{6.29}$$

If (z,ϕ) is admissible and $\rho(z,\phi) = +\infty$, then by definition $z \in L^1(\Omega)$ and $\phi \in W^{1,2}(\Omega)$, but $z \notin \mathrm{BV}(\Omega)$. In this situation (6.29) is trivially satisfied.

The concept of sets of admissible pairs is required since the operator P is discontinuous. For a continuous operator P, we could choose in the above definition $P_{\varepsilon_k} = P$ and, consequently, the set of admissible pairs is $(P(\phi),\phi)$, i.e., single-valued.

In general, the admissible pair according to a function $z \in L^1(\Omega)$ is not unique as the following examples show:

Example 6.9

(i) Let $\phi \in W^{1,2}(\Omega)$ and $\{\varepsilon_k\}$ be a sequence of positive numbers converging to zero, then it follows with (6.25) and Lebesgue's Dominated Convergence Theorem that

$$\lim_{k\to\infty} \|P_{\varepsilon_k}(\phi) - P(\phi)\|_{L^1(\Omega)} = \int_\Omega \lim_{k\to\infty} f_k(t)\,dt = 0\,,$$

where

$$f_k(t) := \begin{cases} 1 + \varepsilon_k^{-1}\phi(t)\,, & -\varepsilon_k < \phi(t) < 0\,,\\ 0\,, & \text{otherwise}\,. \end{cases}$$

By taking $\phi_k = \phi$, it follows with Definition 6.8 (i) that $(P(\phi),\phi)$ is admissible.

(ii) For $\phi \in W^{1,2}(\Omega)$ let $\phi_k := \phi/k$. Moreover, let $\{\varepsilon_k\}$ be a sequence of positive numbers satisfying $k\varepsilon_k \to 0$. Then it follows as above that

$$\lim_{k\to\infty} \|P_{\varepsilon_k}(\phi_k) - P(\phi)\|_{L^1(\Omega)} = \int_\Omega \lim_{k\to\infty} g_k(t)\,dt = 0\,,$$

where

$$g_k(t) := \begin{cases} 1 + (k\varepsilon_k)^{-1}\phi(t)\,, & -k\varepsilon_k < \phi(t) < 0\,,\\ 0\,, & \text{otherwise}\,. \end{cases}$$

This shows that $(P(\phi),0)$ is admissible.

(iii) Let $\kappa \in [0,1]$. Then $(\kappa\chi(\Omega),0)$ is admissible. This follows immediately with Definition 6.8 (i) if we choose $\phi_k := (\kappa-1)/k$ and $\varepsilon_k = 1/k$, since then $P_{\varepsilon_k}(\phi_k) = \kappa$.

In the following lemma we prove that the functional ρ is weakly lower semicontinuous on the set of admissible pairs.

Lemma 6.10 *If there exists a sequence of admissible pairs* $\{(z_k, \phi_k)\}$ *such that* $z_k \to z$ *in* $L^1(\Omega)$ *and* $\phi_k \rightharpoonup \phi$ *in* $W^{1,2}(\Omega)$*, then* (z, ϕ) *is admissible and*

$$\rho(z, \phi) \leq \liminf_{k \in \mathbb{N}} \rho(z_k, \phi_k),$$

where ρ *is defined as in* (6.27).

Proof. First, we prove that (z, ϕ) is admissible. Since (z_k, ϕ_k) is admissible, it follows from Definition 6.8 (i), (iii) that there exist $\varepsilon_k > 0$ and $\psi_k \in W^{1,2}(\Omega)$ such that

$$\|\psi_k - \phi_k\|_{L^2(\Omega)} \leq k^{-1}, \quad \varepsilon_k \leq k^{-1}, \quad \text{and} \quad \|P_{\varepsilon_k}(\psi_k) - z_k\|_{L^1(\Omega)} \leq k^{-1}, \tag{6.30}$$

and, if $\rho(z_k, \phi_k) < \infty$, then also

$$|\rho(z_k, \phi_k) - (\beta\, |P_{\varepsilon_k}(\psi_k)|_{\mathrm{BV}(\Omega)} + \tfrac{1}{2}\, \|\psi_k - \phi_0\|^2_{W^{1,2}(\Omega)})| \leq k^{-1}. \tag{6.31}$$

Since, due to Sobolev's Embedding Theorem, $\phi_k \rightharpoonup \phi$ in $W^{1,2}(\Omega)$ implies that $\phi_k \to \phi$ in $L^2(\Omega)$, we obtain together with $z_k \to z$ in $L^1(\Omega)$ and (6.30) that $\psi_k \to \phi$ in $L^2(\Omega)$ and $P_{\varepsilon_k}(\psi_k) \to z$ in $L^1(\Omega)$. Thus, (z, ϕ) is admissible.

Now (6.27) and (6.31) imply that

$$\begin{aligned} \rho(z, \phi) &\leq \liminf_{k \to \infty} (\beta\, |P_{\varepsilon_k}(\psi_k)|_{\mathrm{BV}(\Omega)} + \tfrac{1}{2}\, \|\psi_k - \phi_0\|^2_{W^{1,2}(\Omega)}) \\ &\leq \liminf_{k \to \infty} (\rho(z_k, \phi_k) + k^{-1}), \end{aligned}$$

which proves the assertion. □

The definition of $\rho(z, \phi)$ is impractical for a numerical realization, since it is defined via a homogenization procedure. The following arguments indicate an explicit characterization of this functional if the curve $\{\phi = 0\}$ is not fat, i.e., the $(n-1)$-dimensional Hausdorff measure of $\{\phi = 0\}$ is finite and $z = \chi_D$, where D has compact support in Ω. In this case, we conjecture that $\partial D = \{\phi = 0\}$ and that

$$\rho(z, \phi) = \beta\, \mathcal{H}^{n-1}(\partial D) + \tfrac{1}{2}\, \|\phi - \phi_0\|^2_{W^{1,2}(\Omega)}.$$

In the case of *curve fattening* the situation is more involved, and we conjecture that

$$\inf_{\{z \,:\, (z,\phi) \text{ is admissible}\}} \rho(z, \phi) = \beta\, \mathcal{H}^{n-1}(\partial D) + \tfrac{1}{2}\, \|\phi - \phi_0\|^2_{W^{1,2}(\Omega)},$$

where ∂D is the minimal surface contained in the set $\{\phi = 0\}$.

In the following proposition we show that there exists a minimal admissible pair (cf. Definition 6.8 (iv)).

Proposition 6.11 *Assume that $F : L^1(\Omega) \to L^2(\tilde{\Omega})$ is continuous and that* (6.3) *has a generalized solution in the sense of Definition* 6.8 *(ii). Then there exists a minimal admissible pair $(z^\dagger, \phi^\dagger)$.*

Proof. Let $\{(z_k, \phi_k)\}$ be a sequence of admissible pairs satisfying $F(z_k) = y$ and $\rho(z_k, \phi_k) \to \rho_{\min}$ as $k \to \infty$, where $\rho_{\min}$ is defined as in (6.28). Note that nothing has to be shown if $\rho_{\min} = \infty$. Therefore, we assume that $\rho_{\min} < \infty$.

According to the definition of admissible pairs, there exist sequences $\{\varepsilon_{k,l}\}$ and $\{\phi_{k,l}\}$ satisfying $P_{\varepsilon_{k,l}}(\phi_{k,l}) \to z_k$ in $L^1(\Omega)$ as $l \to \infty$. This shows that $\|z_k\|_{L^\infty(\Omega)} \le 1$. Since Ω is bounded, we have that $\|z_k\|_{L^1(\Omega)} \le \text{meas}(\Omega)$. From (6.29) it follows that the sequence $\{(z_k, \phi_k)\}$ is uniformly bounded in $\text{BV}(\Omega) \times W^{1,2}(\Omega)$. Consequently, from Sobolev's Embedding Theorem and the fact that $\text{BV}(\Omega)$ is compactly embedded into $L^1(\Omega)$ (see, e.g., [111]), it follows that there exists a convergent subsequence in $L^1(\Omega) \times L^2(\Omega)$, which, for simplicity of notation, we again denote by $\{(z_k, \phi_k)\}$ and the limit is denoted by $(z^\dagger, \phi^\dagger)$. Moreover, $\{\phi_k\}$ is weakly convergent in $W^{1,2}(\Omega)$. It now follows from Lemma 6.10 that $(z^\dagger, \phi^\dagger)$ is admissible and that

$$\rho_{\min} = \lim_{k\to\infty} \rho(z_k, \phi_k) \ge \rho(z^\dagger, \phi^\dagger)\,.$$

Since F is continuous from $L^1(\Omega)$ into $L^2(\tilde{\Omega})$, it follows that $y = F(z_k) \to F(z^\dagger)$ as $k \to \infty$. Now, by Definition 6.8 (iv), $(z^\dagger, \phi^\dagger)$ is a minimal admissible pair. □

Proposition 6.12 *Assume that $F : L^1(\Omega) \to L^2(\tilde{\Omega})$ is continuous. Then $\mathcal{J}_{\lambda,\beta}$ has a generalized minimizer in the sense of Definition* 6.8 *(iii) for all $\lambda, \beta > 0$.*

Proof. Since $(0, 0)$ is an admissible pair, the set of admissible pairs is not empty. Suppose that $\{(z_k, \phi_k)\}$ is a minimizing sequence of admissible pairs, i.e.,

$$0 \le \widetilde{\mathcal{J}}_{\lambda,\beta}(z_k, \phi_k) \to \inf \widetilde{\mathcal{J}}_{\lambda,\beta} \le \widetilde{\mathcal{J}}_{\lambda,\beta}(0, 0) < \infty\,.$$

Using (6.29) and embedding theorems, we can argue as in the previous proof that there exists a convergent subsequence in $L^1(\Omega) \times L^2(\Omega)$, again denoted by $\{(z_k, \phi_k)\}$ with $\{\phi_k\}$ being weakly convergent in $W^{1,2}(\Omega)$. Due to Lemma 6.10, the limit (z, ϕ) is admissible. Moreover, using that ρ is weakly lower semi-continuous it follows that

$$\begin{aligned} \inf \widetilde{\mathcal{J}}_{\lambda,\beta} &= \lim_{k\to\infty} \widetilde{\mathcal{J}}_{\lambda,\beta}(z_k, \phi_k) = \lim_{k\to\infty} \left(\tfrac{1}{2}\lambda \|F(z_k) - y^\delta\|^2_{L^2(\tilde{\Omega})} + \rho(z_k, \phi_k) \right) \\ &\ge \tfrac{1}{2}\lambda \|F(z) - y^\delta\|^2_{L^2(\tilde{\Omega})} + \rho(z, \phi) = \widetilde{\mathcal{J}}_{\lambda,\beta}(z, \phi)\,. \end{aligned}$$

Thus, according to Definition 6.8 (iii), (z, ϕ) is a generalized minimzer of $\mathcal{J}_{\lambda,\beta}$. □

Below, we state a convergence result. The proof uses classical techniques from the analysis of Tikhonov type regularization methods (cf, e.g., [45]). Note that in our functional $\alpha = 1/\lambda$ plays the role of the regularization parameter.

Theorem 6.13 *Assume that $F : L^1(\Omega) \to L^2(\tilde{\Omega})$ is continuous and that a minimal admissible pair $(z^\dagger, \phi^\dagger)$ exists with $\rho(z^\dagger, \phi^\dagger) < \infty$. Moreover, let y^δ satisfy (1.2) and let $\lambda = \lambda(\delta)$ be such that $\lambda(\delta) \to \infty$ and $\delta^2\lambda(\delta) \to 0$ as $\delta \to 0$. Then every sequence $\{(z_k, \phi_k)\}$, where (z_k, ϕ_k) is a generalized minimizer of $\mathcal{J}_{\lambda,\beta}$ for $\lambda = \lambda(\delta_k)$ and $\delta_k \to 0$ as $k \to \infty$, has a convergent subsequence. The limit of every convergent subsequence is a minimal admissible pair.*

Proof. Let $\{\delta_k\}$ be an arbitrary but fixed sequence satisfying $\delta_k \to 0$ as $k \to \infty$. The existence of a generalized minimizer (z_k, ϕ_k) of $\mathcal{J}_{\lambda,\beta}$ for $\lambda = \lambda(\delta_k)$ follows from Proposition 6.12. By definition of these minimizers (see Definition 6.8 (iii)),

$$\tfrac{1}{2}\lambda(\delta_k)\,\|F(z_k) - y_k^\delta\|_{L^2(\tilde{\Omega})}^2 + \rho(z_k, \phi_k) \le \tfrac{1}{2}\lambda(\delta_k)\delta_k^2 + \rho(z^\dagger, \phi^\dagger)\,.$$

Consequently,

$$\lim_{k\to\infty} F(z_k) = y \tag{6.32}$$

and

$$\limsup_{k\to\infty} \rho(z_k, \phi_k) \le \rho(z^\dagger, \phi^\dagger)\,. \tag{6.33}$$

Using (6.29) and embedding theorems, we can argue as in the proof of Proposition 6.11 that there exists a convergent subsequence in $L^1(\Omega) \times L^2(\Omega)$, again denoted by $\{(z_k, \phi_k)\}$, with $\{\phi_k\}$ being weakly convergent in $W^{1,2}(\Omega)$. Due to Lemma 6.10, the limit (z, ϕ) is admissible. Moreover, using that ρ is weakly lower semicontinuous it follows together with (6.33) that

$$\rho(z, \phi) \le \liminf_{k\to\infty} \rho(z_k, \phi_k) \le \limsup_{k\to\infty} \rho(z_k, \phi_k) \le \rho(z^\dagger, \phi^\dagger)\,.$$

This, together with (6.32) and Definition 6.8 (iv) shows that (z, ϕ) is a minimal admissible pair. □

To avoid the calculation of the infimum in (6.27), for numerical computations we will approximate the regularization functional in Definition 6.8 (iii) by the following stabilized functional

$$\mathcal{J}_{\lambda,\beta}^{\varepsilon}(\phi) := \tfrac{1}{2}\lambda\|F(P_\varepsilon(\phi)) - y^\delta\|_{L^2(\tilde{\Omega})}^2 + \beta\,|P_\varepsilon(\phi)|_{\mathrm{BV}(\Omega)} + \tfrac{1}{2}\,\|\phi - \phi_0\|_{W^{1,2}(\Omega)}^2\,.$$

Proposition 6.14 *Assume $F : L^1(\Omega) \to L^2(\tilde{\Omega})$ is continuous. Then, $\mathcal{J}_{\lambda,\beta}^{\varepsilon}$ has a minimizer for every $\varepsilon, \lambda, \beta > 0$.*

Proof. The proof is similar to the proof of Proposition 6.11. □

In the following proposition we show that for $\varepsilon \to 0$ the minimizers of $\mathcal{J}^{\varepsilon}_{\lambda,\beta}$ approximate generalized minimizers of $\mathcal{J}_{\lambda,\beta}$. This justifies the use of the functional $\mathcal{J}^{\varepsilon}_{\lambda,\beta}$ for a numerical realization. Note that in contrast to the minimizer of $\mathcal{J}^{\varepsilon}_{\lambda,\beta}$, which is a function of $W^{1,2}(\Omega)$, the minimizer of $\widetilde{\mathcal{J}}_{\lambda,\beta}$ is an admissible pair $(z_{\lambda,\beta}, \phi_{\lambda,\beta})$. Recall that the function $z_{\lambda,\beta}$ is not uniquely defined by $\phi_{\lambda,\beta}$ if it attains critical values in a neighbourhood of the zero level set.

Proposition 6.15 *Assume that $F : L^1(\Omega) \to L^2(\tilde{\Omega})$ is continuous and that $\lambda, \beta > 0$ are fixed. Then*

$$\liminf_{\varepsilon \to 0} \mathcal{J}^{\varepsilon}_{\lambda,\beta}(\phi^{\varepsilon}_{\lambda,\beta}) = \inf \widetilde{\mathcal{J}}_{\lambda,\beta}\,,$$

where $\phi^{\varepsilon}_{\lambda,\beta}$ is a minimizer of $\mathcal{J}^{\varepsilon}_{\lambda,\beta}$ and $\widetilde{\mathcal{J}}_{\lambda,\beta}$ is defined as in (6.26). *Moreover, there exists a sequence $\{\varepsilon_k\}$ with $\varepsilon_k \to 0$ as $k \to \infty$ such that $(P_{\varepsilon_k}(\phi^{\varepsilon_k}_{\lambda,\beta}), \phi^{\varepsilon_k}_{\lambda,\beta})$ converges to a generalized minimizer of $\mathcal{J}_{\lambda,\beta}$, i.e., a minimzer of $\widetilde{\mathcal{J}}_{\lambda,\beta}$.*

Proof. Let ε_k be an arbitrary sequence of positive numbers converging to zero. Since, due to Proposition 6.14, $\mathcal{J}^{\varepsilon_k}_{\lambda,\beta}$ has a minimizer $\phi^{\varepsilon_k}_{\lambda,\beta}$, and since, due to (6.25),

$$\begin{aligned}
\mathcal{J}^{\varepsilon_k}_{\lambda,\beta}(\phi^{\varepsilon_k}_{\lambda,\beta}) &= \tfrac{1}{2}\lambda \|F(P_{\varepsilon_k}(\phi^{\varepsilon_k}_{\lambda,\beta})) - y^\delta\|^2_{L^2(\tilde{\Omega})} + \beta\, |P_{\varepsilon_k}(\phi^{\varepsilon_k}_{\lambda,\beta})|_{\mathrm{BV}(\Omega)} \\
&\quad + \tfrac{1}{2}\|\phi^{\varepsilon_k}_{\lambda,\beta} - \phi_0\|^2_{W^{1,2}(\Omega)} \\
&\leq \tfrac{1}{2}\lambda \|F(0) - y^\delta\|^2_{L^2(\tilde{\Omega})} + \tfrac{1}{2}\|\varepsilon_k \chi(\Omega) + \phi_0\|^2_{W^{1,2}(\Omega)} \\
&= \mathcal{J}^{\varepsilon_k}_{\lambda,\beta}(-\varepsilon_k \chi(\Omega))\,,
\end{aligned}$$

the sequence $\{(P_{\varepsilon_k}(\phi^{\varepsilon_k}_{\lambda,\beta}), \phi^{\varepsilon_k}_{\lambda,\beta})\}$ is uniformly bounded in $\mathrm{BV}(\Omega) \times W^{1,2}(\Omega)$. Thus, using embedding theorems, we can argue as in the proof of Proposition 6.11 that there exists a convergent subsequence in $L^1(\Omega) \times L^2(\Omega)$, again denoted by $\{(P_{\varepsilon_k}(\phi^{\varepsilon_k}_{\lambda,\beta}), \phi^{\varepsilon_k}_{\lambda,\beta})\}$, with $\{\phi^{\varepsilon_k}_{\lambda,\beta}\}$ being weakly convergent in $W^{1,2}(\Omega)$. According to Definition 6.8 (i), the limit $(z, \phi) \in L^1(\Omega) \times W^{1,2}(\Omega)$ is admissible. It now follows with the continuity of F and the definition of $\widetilde{\mathcal{J}}_{\lambda,\beta}$ (see (6.26)) that

$$\widetilde{\mathcal{J}}_{\lambda,\beta}(z, \phi) \leq \lim_{k \to \infty} \mathcal{J}^{\varepsilon_k}_{\lambda,\beta}(\phi^{\varepsilon_k}_{\lambda,\beta}) \tag{6.34}$$

and hence

$$\inf \widetilde{\mathcal{J}}_{\lambda,\beta} \leq \liminf_{\varepsilon \to 0} \mathcal{J}^{\varepsilon}_{\lambda,\beta}(\phi^{\varepsilon}_{\lambda,\beta})\,.$$

According to Proposition 6.12, there exists a minimizing pair $(z_{\lambda,\beta}, \phi_{\lambda,\beta})$ of $\widetilde{\mathcal{J}}_{\lambda,\beta}$. Taking into account the definition of admissible pairs, there exists a sequence

$\{\varepsilon_k\}$ of positive numbers converging to zero and a sequence $\{\phi_k\}$ in $W^{1,2}(\Omega)$ satisfying

$$\begin{aligned} (P_{\varepsilon_k}(\phi_k), \phi_k) &\rightarrow (z_{\lambda,\beta}, \phi_{\lambda,\beta}) \quad \text{in} \quad L^1(\Omega) \times L^2(\Omega)\,, \\ \rho(z_{\lambda,\beta}, \phi_{\lambda,\beta}) &= \lim_{k\to\infty} \{\beta\,|P_{\varepsilon_k}(\phi_k)|_{\mathrm{BV}(\Omega)} + \|\phi_k - \phi_0\|^2_{W^{1,2}(\Omega)}\}\,. \end{aligned} \tag{6.35}$$

For this sequence $\{\varepsilon_k\}$, a subsequence (again denoted by $\{\varepsilon_k\}$) and (z, ϕ) exist as above satisfying (6.34). This yields the estimate

$$\begin{aligned} \widetilde{\mathcal{J}}_{\lambda,\beta}(z,\phi) &\le \liminf_{k\to\infty} \mathcal{J}^{\varepsilon_k}_{\lambda,\beta}(\phi^{\varepsilon_k}_{\lambda,\beta}) \le \liminf_{k\to\infty} \mathcal{J}^{\varepsilon_k}_{\lambda,\beta}(\phi_k) \\ &= \tfrac{1}{2}\lambda\|F(z_{\lambda,\beta}) - y^\delta\|^2_{L^2(\tilde{\Omega})} + \rho(z_{\lambda,\beta}, \phi_{\lambda,\beta}) = \inf \widetilde{\mathcal{J}}_{\lambda,\beta}\,. \end{aligned}$$

This proves the assertions. □

It follows from the proof above that for every sequence $\{\varepsilon_k\}$ having the property (6.35) a subsequence exists satisfying the convergence result of Proposition 6.15.

Remark 6.16 In this chapter we have discussed iterative methods for solving level set equations of the form (6.3) (cf. (6.7), (6.23)). In comparison with gradient flow methods, Newton type methods (such as (6.23)) are more natural for solving (6.3), since the operator P is discontinuous and, in general, the subdifferential of the energy functional

$$\tfrac{1}{2}\lambda\|F(z) - y^\delta\|^2_{L^2(\tilde{\Omega})} + \beta\,|z|_{\mathrm{BV}(\Omega)}$$

is set-valued, which requires an additional selection step for the gradient flow methods.

A Newton type method for minimizing the energy function $\mathcal{J}_{\lambda,\beta}$ (see (6.24)) can be interpreted as realization of an implicit time step of the Euler equation of the time discretized time evolution process. From a theoretical point of view the existence of a minimizer is guaranteed by our analysis.

7 Applications

Iterative methods have been used for solving inverse problems such as scattering problems (see, e.g., [66]). A vast amount of shape recovery problems (which are by nature non-linear) in scattering and electrical impedance imaging have been solved with iterative regularization methods. We refer to [72, 73, 74, 76, 103] and the references therein. Later on, these methods have been supplemented by level set techniques for shape recovery (see [25]), which share the same objective but allow for multiple connected objects. For numerical realization both of them essentially use the same techniques of shape derivatives, but the numerical realization is different.

In this book we present two applications. The first one is from Schlieren tomography. It is solved by a Landweber–Kaczmarz method, which we consider a prime candidate for solving tomographic problems. The second example concerns a parameter estimation problem from nonlinear magnetics. It is solved by various iterative methods which have been analyzed in this book. Moreover, the numerical performance of the algorithms is compared.

7.1 Reconstruction of transducer pressure fields from Schlieren images

We deal with the following problem from Schlieren tomography: in order to test and improve ultrasound transducers it is necessary to estimate their unperturbed pressure fields. Nowadays this is done as follows: an ultrasound transducer is placed on the top of a water tank and sends pressure waves into the tank. The generated pressure variations within the water tank cause density changes that can scatter light. At an appropriate time instance the light intensity of a laser beam is measured on one side of the tank via an optical Schlieren system (see Figure 7.1). This procedure is repeated for a set of rotation angles of the transducer. For practical aspects on the realization of Schlieren systems and more background information see, e.g., [26, 131].

This problem may be modelled as follows: let $D \subset \mathbb{R}^2$ denote the unit disc and let $\sigma_i \in \mathcal{S}^1$, $i = 0, \dots, N-1$, be a set of recording angles. (Associated with each σ_i there is an angle $\varphi_i \in [0, \pi)$ with $\sigma_i = (\cos\varphi_i, \sin\varphi_i)$). The Schlieren system output in direction of σ_i is then given via the square of the Radon transform, i.e.,

$$F_i(p) := R_i^2(p) \quad \text{with} \quad R_i(p)(s) := \int_{\mathbb{R}} p(s\sigma_i + r\sigma_i^{\perp})\, dr\,, \quad s \in [-1, 1]\,,$$

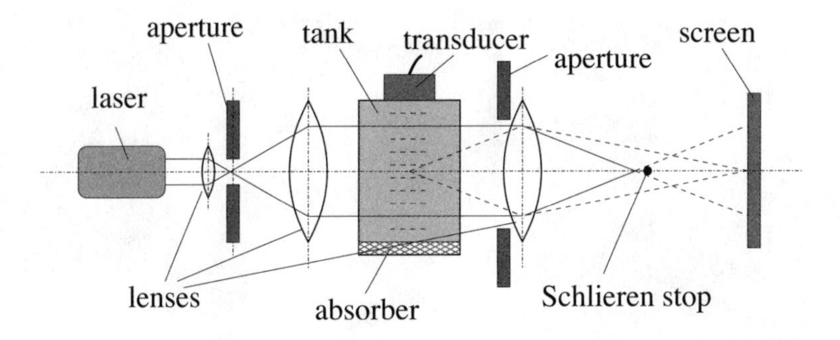

Figure 7.1 A Schlieren optical system

where s is the signed distance of the line $L_i(s) := s\sigma_i + \mathbb{R}\sigma_i^{\perp}$ from the origin in D and p denotes the pressure. Note that, since in general the pressure function has positive and negative values, the Schlieren data cannot be reduced to standard x-ray tomography. Now, the problem of reconstructing the pressure function p consists in solving the equation

$$F(p) := (F_0(p), \ldots, F_{N-1}(p)) = y^\delta = (y_0^\delta, \ldots, y_{N-1}^\delta)\,.$$

It was shown in [61] that the operators F_i are continuous and Fréchet-differentiable from $L^4(D)$ to $L^2[-1,1]$ and hence, due to the embedding theorem, also from $H_0^1(D)$ to $L^2[-1,1]$. Since we are interested in functions p with jumps, we choose a different space: in our numerical computations we approximate the pressure function by piecewise constant functions on a uniform 480×480 grid of $[-1,1]^2$ assuming that the function is 0 outside of D. Therefore, we define the operators F_i on the appropriate finite-dimensional space $\mathcal{X}$ equipped with the L^2-norm into $L^2[-1,1]$. The operators F_i are then continuous and Fréchet-differentiable even on $\mathcal{X}$ with

$$F_i'(p)h = 2R_i(p)R_i(h) \qquad \text{and} \qquad F_i'(p)^*v = 2PR_i^*(R_i(p)v)\,,$$

where $R_i^* z(\xi) = z(\langle\, \xi, \sigma_i \,\rangle)$ and P is the orthogonal projector from $L^2(D)$ onto $\mathcal{X}$. Since the grid is very fine, Pz may be approximated well enough by the piecewise constant spline interpolant of z.

The synthetic Schlieren data were simulated for 251 different equally distributed angles within the interval $[0,\pi)$. Uniformly distributed noise of level $\delta = 0.01\%$ was added to the synthetic data. All numerical simulations were performed by Richard Kowar (University of Innsbruck) on an Intel Xenon CPU 5160 with 3 GHz. The basic system software was Fedore core FC6. We compare the performance of the following three methods: the Landweber–Kaczmarz method (3.85) (LK), its modification (3.88) (mLK), and the Levenberg–Marquardt method (4.2)

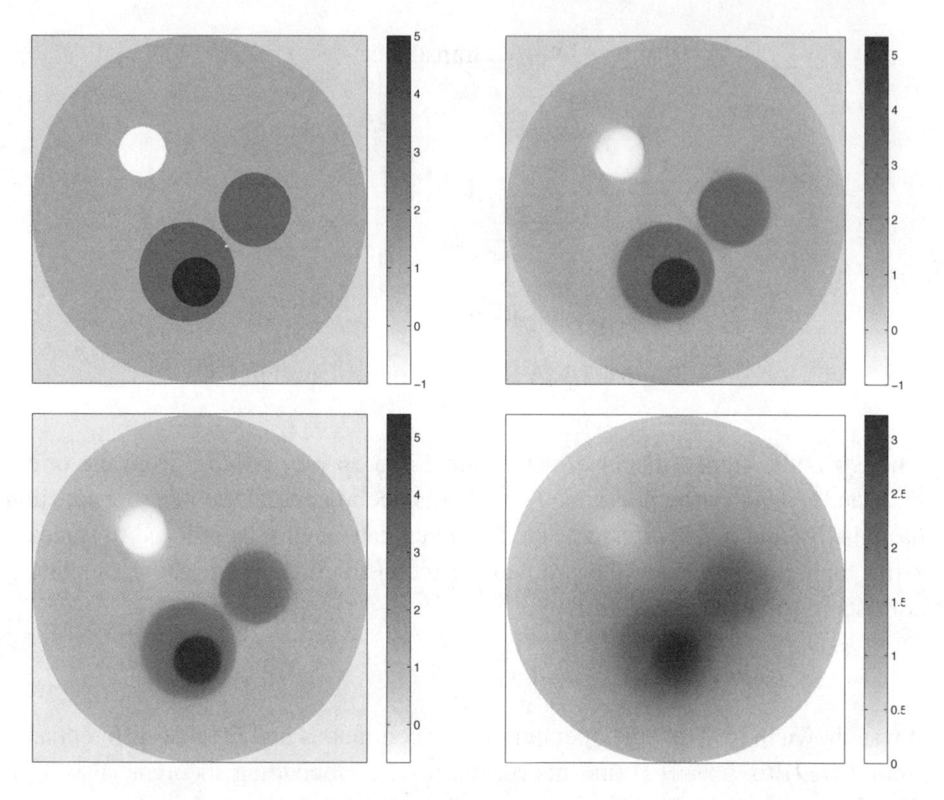

Figure 7.2 Example 1: exact function (upper left), mLK (upper right), LK (lower left), LM (lower right)

(LM), where the linear system appearing in the latter was solved approximately by a few conjugate gradient steps. In the Landweber–Kaczmarz method and its modification a scaling parameter was introduced similarly to (2.24) to guarantee the scaling property (2.3).

Our numerical simulations demonstrate that for this problem the modified Landweber–Kaczmarz method is the fastest method and yields the best results.

We present two examples of synthetic pressure functions. Both are piecewise continuous, however, the second one has a large area of zero pressure that contains several stripes, where then the derivative operators F_i' (with $L_i(s)$ subset of the stripes) vanish. It turns out that such pressure functions are much more difficult to estimate than others.

For the results of the first example see Figure 7.2. The reconstruction quality with mLK and with LK are equally good and the reconstruction with LM yields a strongly smoothed version of the exact pressure function. The estimated negative part in the LM-reconstuction is positive and almost equal to zero. τ in the discrep-

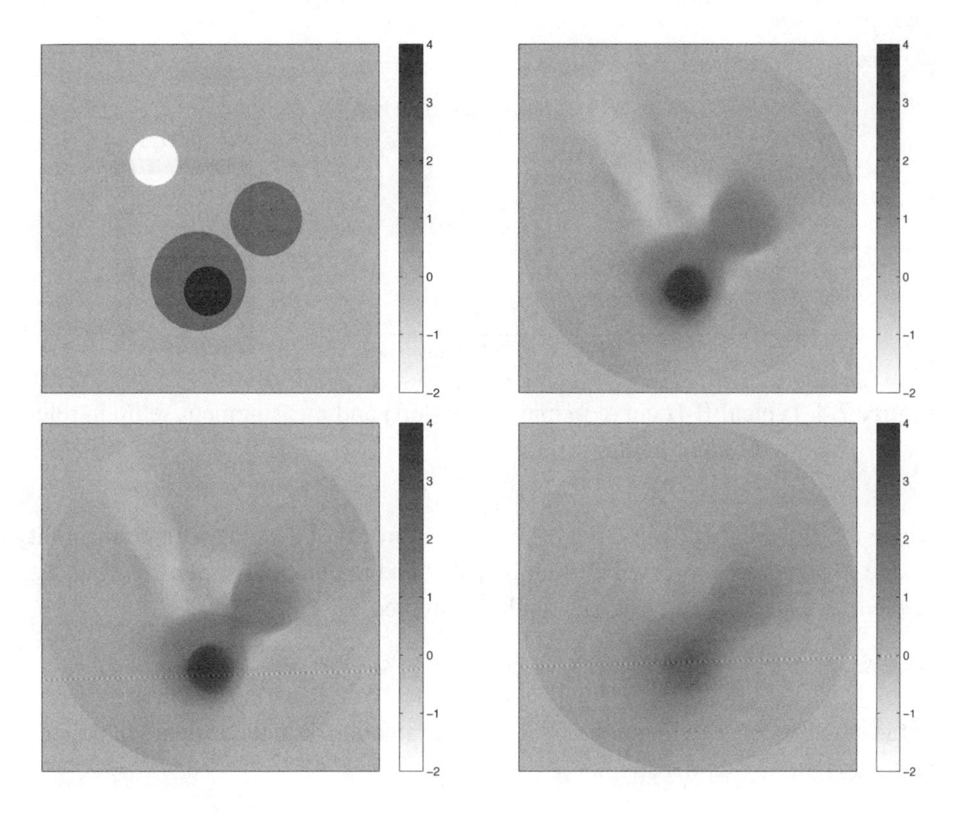

Figure 7.3 Example 2: exact function (upper left), mLK (upper right), LK (lower left), LM (lower right)

ancy principle was chosen to be 4.7. The CPU times were 75 s for mLK, 163 s for LK, and 102 s for LM.

The results of the second example are given in Figure 7.3. The reconstruction quality with mLK and with LK are again equal, whereas the reconstruction with LM is much worse. One can also see that the reconstructions via mLK and LK contain negative artifacts, $\tau = 3.5$. The CPU times were 125 s for mLK, 147 s for LK, and 186 s for LM.

7.2 Identification of B-H curves in nonlinear magnetics

Magnetostatic fields can be described by a subset of Maxwell's equations leading to the following system of PDEs for the magnetic vector potential $\mathbf{A}$,

$$\nabla \times (\nu \nabla \times \mathbf{A}) = \mathbf{J}_{\text{imp}}\,, \tag{7.1}$$

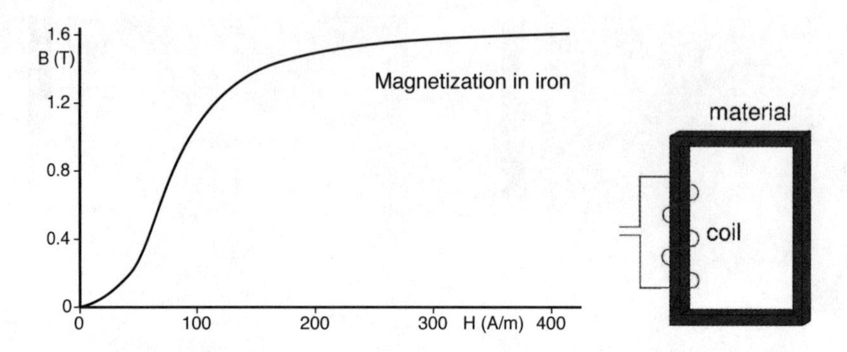

Figure 7.4 Typical B-H curve in magnetics (left) and measurement setup in reluctivity identification (right)

where ν is the magnetic reluctivity (i.e., the reciprocal of the magnetic permeability μ) and $\mathbf{J}_{\text{imp}}$ is the impressed current density. The magnetic flux density $\mathbf{B}$ as well as the magnetic field intensity $\mathbf{H}$ can be expressed as

$$\mathbf{B} = \nabla \times \mathbf{A}\,, \qquad \mathbf{H} = \nu \mathbf{B}\,.$$

In the situation of high magnetic fields, the parameter ν is not constant but depends on the magnetic flux density, i.e.,

$$H = \nu(B)B\,, \tag{7.2}$$

where H and B denote the magnitude of the vectorial quantities $\mathbf{H}$ and $\mathbf{B}$, respectively.

It is a practically relevant inverse problem in material characterization to determine the curve $B \mapsto \nu(B)$ or equivalently the so-called B-H curve from indirect measurements (cf. [90], see Figure 7.4 for a typical example).

In a usual experimental setup for determining the reluctivity ν in (7.2) (see Figure 7.4), the impressed current density takes the form

$$\mathbf{J}_{\text{imp}} = \begin{cases} \dfrac{I}{|\Omega_c|}\mathbf{e}_J\,, & \text{in } D_c\,, \\ 0\,, & \text{else}\,, \end{cases}$$

and additional measurements of the magnetic flux through the coil

$$\Phi = \oint_{C_c} \mathbf{A} \cdot ds = \int_{\Omega_c} \mathbf{B} \cdot \mathbf{n}\, d\sigma \tag{7.3}$$

are available. Here, I is the prescribed current, $\mathbf{e}_J$ is the unit vector in current direction, D_c is the region covered by the excitation coil, Ω_c is its cross sectional

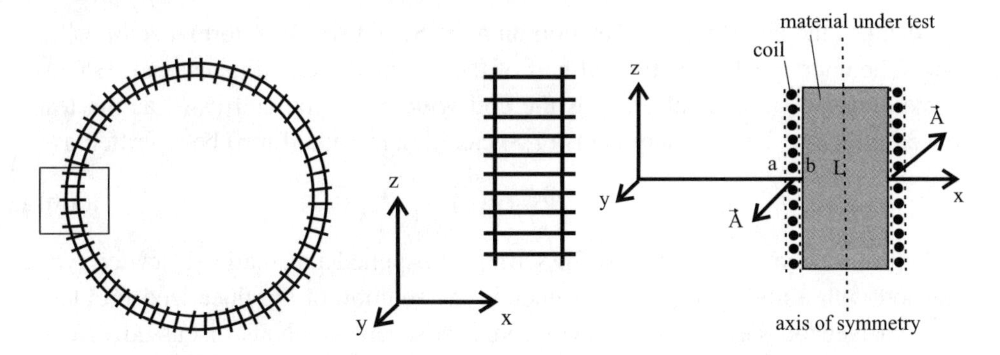

Figure 7.5 Schematic of probe with coil (left), quasi straight detail (middle), and cut along x-z plane (right)

area, and C_c is the boundary curve of Ω_c. Note that the identity of the integrals in (7.3) holds due to Stokes' Theorem. The computational domain comprises the probe, the coil, as well as the surrounding air region. Conditions on the outer boundary are

$$\mathbf{n} \times \mathbf{A} = 0\,. \tag{7.4}$$

Therefore, the inverse problem under consideration is the identification of the reluctivity $\nu = \nu(B)$ in the nonlinear B-H relation (7.2) from measurements of the magnetic flux Φ for different currents I in an excitation coil.

By an appropriate experimental setup, this can be reduced to the spatially 1D case: consider a ring-shaped probe entwined with an excitation/measurement coil according to Figure 7.5 with a large interior radius so that the curvature can be neglected and the magnetic flux density points into z direction but does not vary in z direction. Moreover, we consider a cut along the x-z plane in which $\mathbf{A}$ must be parallel to the y axis and dependence of $\mathbf{A}$ and $\mathbf{B}$ on y may be neglected, so that altogether

$$\mathbf{B}(x,y,z) = (0,0,B^z(x))^T$$

or equivalently

$$\mathbf{A}(x,y,z) = (0,A^y(x),0)^T.$$

With $u(x) := A^y(x)$ the system (7.1) with boundary conditions (7.4) becomes

$$\begin{aligned} -c(x,u_x(x))_x &= I\chi_{[a,b]}(x)\,, \qquad x \in (0,L)\,, \\ u(0) = u(L) &= 0\,, \end{aligned} \tag{7.5}$$

where

$$c(x,\zeta) = \begin{cases} \mu_0^{-1}\zeta\,, & x \in [0,b)\,, \\ \nu(\zeta)\zeta\,, & x \in [b,L]\,, \end{cases}$$

χ_S denotes the characteristic function on a set S, and we have normalized $|\Omega_c|$ to one. The interval $(0, L)$ is the left half of the symmetric one-dimensional domain containing the subinterval $[a, b]$ as the coil winding region and $[b, L]$ as the test material region. The measurements (7.3) can (in a rescaled form) be rewritten as

$$\Phi = u(a)\,. \tag{7.6}$$

In [90] we considered the fully 3D setting and applied a regularized Newton type method with a multigrid preconditioner in the solution of the linearized problem in each Newton step. Here we will compare several methods discussed in this book. For this purpose, we restrict ourselves to the one-dimensional setting (7.5), (7.6) to avoid technicalities arising from the appropriate spline representation of the B-H curves (cf., e.g., [128] for a recent paper on this subject) as well as the numerical solution of the 3D Maxwell system and refer to [90] for further details and references. Moreover we wish to point the reader to [1] for an application of our multigrid method to a large scale inverse problem in pollution detection.

Our aim is to determine the function $f : [\underline{\lambda}, \overline{\lambda}] \to \mathbb{R}$ defined by

$$f^{-1}(\zeta) = \nu(\zeta)\zeta\,,$$

i.e., f is the function whose graph describes the B-H curve. For simplicity we consider a fixed interval $[\underline{\lambda}, \overline{\lambda}]$ and refer to [109] for the identification of a nonlinearity in a PDE as well as its domain. With the forward operator

$$\begin{aligned} F : \mathcal{D}(F) \subseteq H^2(\underline{\lambda}, \overline{\lambda}) \quad &\to \quad L^2([\underline{I}, \overline{I}]) \\ f \quad &\mapsto \quad (I \mapsto u^I(a)) \quad \text{where } u^I \text{ solves (7.5)} \end{aligned}$$

and the data $y = (I \mapsto \Phi)$, this can be written as an operator equation $F(f) = y$. The choice of $L^2([\underline{I}, \overline{I}])$ as our data space is motivated by the fact that we can only measure point but not derivative values and the requirement of working with a Hilbert space. The domain of F is chosen as

$$\mathcal{D}(F) = \{f \in H^2(\underline{\lambda}, \overline{\lambda}) \;:\; f'(\lambda) \geq \mu_0 > 0\}$$

to guarantee well-definedness of the forward problem. Note that by the continuity of the embedding $H^2(\underline{\lambda}, \overline{\lambda}) \to C^1([\underline{\lambda}, \overline{\lambda}])$, the domain $\mathcal{D}(F)$ has nonempty interior. In our discretization of the problem we use an ansatz for f' (rather than for f) by piecewise linear splines to easily monitor monotonicity of f by positivity of the coefficients of f' in terms of the hat function basis. We mention in passing that with this ansatz one can as well easily safeguard monotonicity by projection (cf. [91]), which did not appear to be necessary in our example, though. From f' the original function f is obtained by (analytical) integration using the fact that $f(0) = 0$.

The spaces Y_l for discretization in image space are defined as continuous piecewise linear functions on a grid of size h_l, i.e., Y_l is the span of the hat functions on an equidistant partition of $[\underline{I}, \overline{I}]$ with mesh size

$$h_l = 2^{-l}.$$

It was shown in [90] that within the natural setting described above the inverse problem of B-H curve identification is as ill-posed as $3/2$ differentiations. Hence the parameter ξ_l in the smoothing step of the multigrid method should be chosen according to

$$\xi_l \sim h_l^{-3}.$$

To satisfy condition (5.13) with a maximal ξ_l in order to obtain the best possible convergence rate, we approximately compute $\overline{\gamma}_{l-1}$, i.e., the maximal singular value of the operator $(I - Q_{l-1})K_\sharp$, which can be done with small additional effort, e.g., by a few steps of the power method, and set $\xi_l = 1/\overline{\gamma}_{l-1}^2$.

Figure 7.6 shows the results obtained with 1000 Landweber iterations (LW), 1000 steepest descent (SD) iterations, 1000 minimal error iterations (ME), 100 steps of the Levenberg–Marquardt method (LM), 30 steps of the (IRGN), and with 6 levels of the nonlinear full multigrid method (NFMGM) for noiseless data.

All calculations were done on a laptop with an Intel Pentium M 1.6 GHz processor using Matlab 7.2. The CPU times and errors as well as residuals are displayed in the table below, where f_k denotes the result of the respective method:

	CPU time (s)	$\frac{\|f_k - f\|_{H^2}}{\|f\|_{H^2}}$	$\frac{\|F(f_k) - y\|_{L^2}}{\|y\|_{L^2}}$
LW (1000 steps)	3150	0.1016	0.0066
SD (1000 steps)	3148	0.0881	0.0050
ME (1000 steps)	3081	0.0818	0.0041
LM (100 steps)	326	0.0505	0.0019
IRGN (30 steps)	97	0.0264	0.0040
NFMGM (6 levels)	88	0.0236	0.0027

In Figure 7.7 we show a comparison of the Levenberg–Marquardt method and the iteratively regularized Gauss–Newton method after five iterations. It seems that LM is better at the beginning whereas from Figure 7.6 it can be seen that while IRGN yields good results, LM appears to get the same difficulties with reconstructing f close to the right hand boundary as the Landweber type iterations. This is possibly due to the fact that the initial guess appearing in each of the IRGN steps helps to prevent the slope at the right hand boundary point from differing too much from the initial slope at this point.

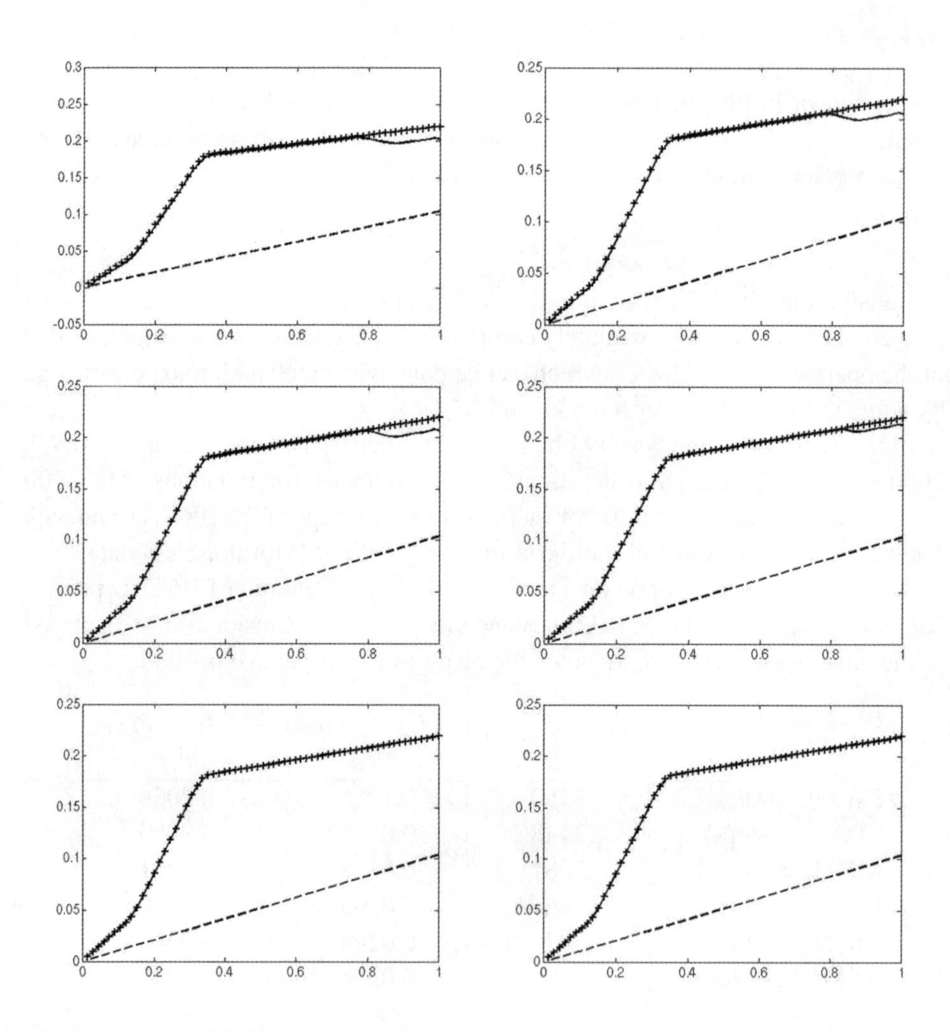

Figure 7.6 Results (solid line) for 1000 LW (upper left), 1000 SD (upper right), 1000 ME (center left), 100 LM (center right), 30 IRGN (lower left), and 6 NFMGM (lower right); starting value: dashed, exact solution: pluses

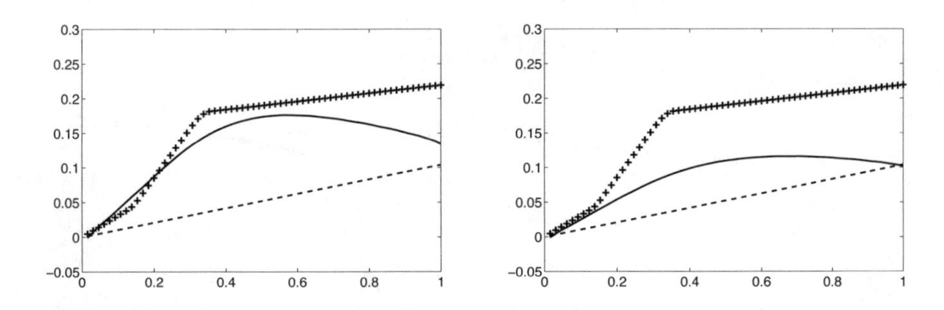

Figure 7.7 Fifth iterate of the LM (left) and of the IRGN (right), respectively; starting value: dashed, exact solution: pluses

Figure 7.8 shows intermediate results for the nonlinear full multigrid method. Obviously, a considerable portion of the solution is already recovered at relatively coarse levels of discretization, so that only few of the more costly iterations on the finer levels are necessary, which explains the computational efficiency of this method.

To test the performance of the nonlinear full multigrid method with noisy data, we added uniformly distributed random noise to the synthetic data, that we had generated on a finer grid than the one used in computations to avoid an inverse crime. As a stopping rule for determining the discretization level in dependence of the noise level, we used the discrepancy principle with $\tau = 2$. Results at a noise level of 1% and 10% are displayed in Figure 7.9. The convergence behaviour for this example as the noise level tends to zero is shown in the table below:

δ	l_*	$\frac{\|f_{l_*}^\delta - f\|_{H^2}}{\|f\|_{H^2}}$
8%	1	0.3515
4%	2	0.2354
2%	3	0.1783
1%	4	0.1165
0.5%	5	0.0752
0.25%	6	0.0517

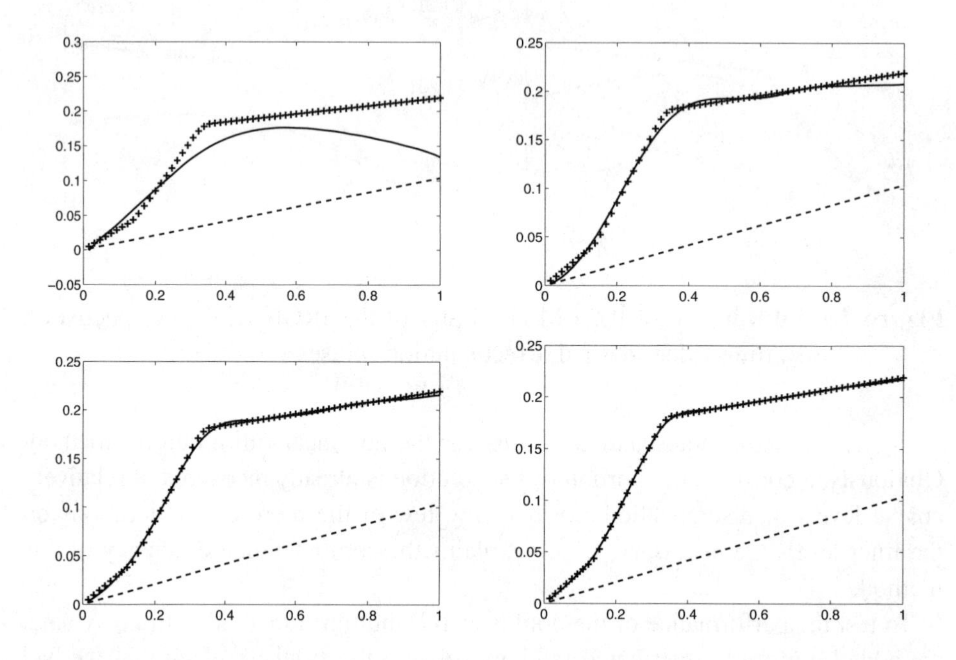

Figure 7.8 First (upper left), second (upper right), third (lower left), and fourth (lower right) level of the NFMGM; starting value: dashed, exact solution: pluses

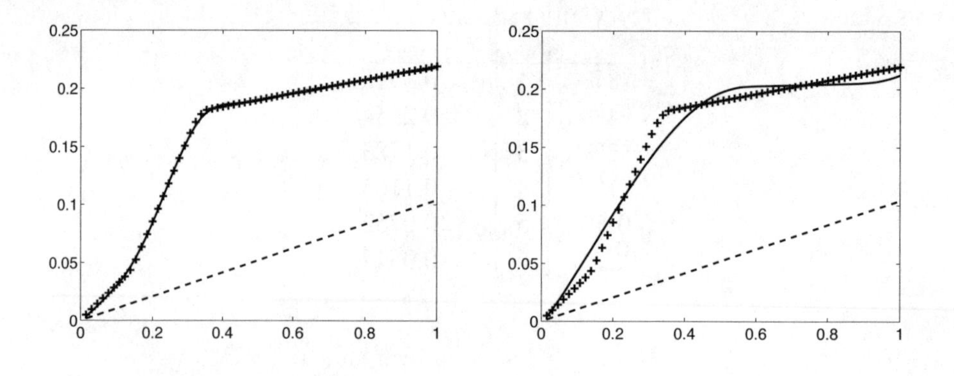

Figure 7.9 Result (solid line) of nonlinear full multigrid with 1% (left) and 10% (right) noise in the data; starting value: dashed, exact solution: pluses

8 Comments

In this chapter we give comments on some other iterative regularization approaches that are not covered in this book.

Over the years there have been developed various methods for solving ill-posed and inverse problems in a stable way.

We have already mentioned Tikhonov regularization, where the operator F is a mapping between Hilbert spaces $\mathcal{X}$ and $\mathcal{Y}$ and a solution of (1.1) is approximated by the minimizer of the Tikhonov functional defined in (1.3). Recently (see, e.g., [145]), there have been many studies devoted to variational regularization in Banach spaces, consisting in minimizing the functional

$$x \mapsto G(\|F(x) - y^\delta\|) + \alpha\Phi(\|x - x_0\|)\,,$$

where G and Φ are typically primitives of weight functions (see [31] for more background on weight functions and their role in convex analysis).

This book focuses on solving operator equations (1.1) in Hilbert spaces with iterative regularization methods in a stable way by *early termination* of the iteration. These methods can be similarly defined in a Banach space setting. To see this, let $F : \mathcal{X} \to \mathcal{Y}$ be an operator between Banach spaces $\mathcal{X}$ and $\mathcal{Y}$, and take into account that the steepest descent direction of the functional

$$x \mapsto \tfrac{1}{2}\|F(x) - y^\delta\|^2$$

is given by $-F'(x)^{\#}(F(x) - y^\delta)$, where $F'(x)^{\#} : \mathcal{Y}^* \to \mathcal{X}^*$ is the dual adjoint operator. Note that the operator is defined on the dual spaces of the Banach spaces $\mathcal{X}$ and $\mathcal{Y}$, respectively. Then for instance the Landweber iteration reads as follows

$$\mathcal{J}_{\mathcal{X}}(x_{k+1}^\delta - x_k^\delta) = -F'(x_k^\delta)^{\#}(F(x_k^\delta) - y^\delta)\,, \tag{8.1}$$

where $\mathcal{J}_{\mathcal{X}}$ is the duality mapping from $\mathcal{X}$ into $\mathcal{X}^*$ (see [31] for more background on duality mappings).

We emphasize that iterative methods as studied in this book are based on the least squares fit for minimizing $\frac{1}{2}\|F(x) - y^\delta\|^2$. Iterative methods can also be based on other fit-to-data functionals: for instance consider the minimization of

$$x \mapsto G(\|F(x) - y^\delta\|)\,,$$

with a convex function G, then the Landweber iteration iteration reads as follows

$$\mathcal{J}_{\mathcal{X}}(x_{k+1}^\delta - x_k^\delta) = -F'(x_k^\delta)^{\#}G'(F(x_k^\delta) - y^\delta)\,,$$

where the evaluation of the function G' is understood pointwise.

In Banach spaces, a similar approach to (8.1) for solving ill-posed problems has been suggested in [146], where the iteration

$$\mathcal{J}_{\mathcal{X}}(x_{k+1}^\delta) - \mathcal{J}_{\mathcal{X}}(x_k^\delta) = -F'(x_k^\delta)^{\#}(F(x_k^\delta) - y^\delta) \tag{8.2}$$

has been considered. Noting that $\mathcal{J}_{\mathcal{X}}(x - x_k^\delta)$ is the steepest descent functional of $\frac{1}{2}\|x - x_k^\delta\|^2$, while $\mathcal{J}_{\mathcal{X}}(x) - \mathcal{J}_{\mathcal{X}}(x_k^\delta)$ is the steepest descent functional of $\frac{1}{2}\|x\|^2 - \frac{1}{2}\|x_k^\delta\|^2 - \langle \mathcal{J}_{\mathcal{X}}(x_k^\delta), x \rangle$, which is the Bregman distance, the approach in (8.1) could be called explicit Tikhonov-Morozov iteration whereas the one in (8.2) could be called explicit Bregman iteration. Bregman distance regularization methods that are implicit variants of the methods proposed in [146] have been considered for instance in [21].

Another research topic on iterative methods concerns the minimization of *penalized* functionals

$$x \mapsto \|F(x) - y^\delta\|^2 + \alpha \|x\|_{\mathcal{X}}, \tag{8.3}$$

where the penalization term $\|\cdot\|_{\mathcal{X}}$ (with $\mathcal{X}$ a Banach space) is used to take into account additional constraints, such as sparsity. These iteration methods yield approximations for minimizers of the functional above and not of the original problem (1.1). Typical results on convergence of the iterates are shown for $k \to \infty$ for fixed $\alpha > 0$ (see [33]). Since the minimization of the functional in (8.3) is a well-posed problem, it is not necessary to terminate the iteration number k as $\delta \to 0$.

In case of an L^1-penalization, the following semi-implicit fixed point iteration

$$x_{k+1}^\delta = x_k^\delta - F'(x_k^\delta)^*(F(x_k^\delta) - y^\delta) - \alpha\,\mathrm{sgn}(x_{k+1}^\delta)$$

was considered in [33]. Here, sgn is a set-valued function, defined as 1 for positive values, -1 for negative values, and $\mathrm{sgn}(0) \in [-1, 1]$. Note that the operator $x + \alpha\,\mathrm{sgn}(x)$ is invertible.

As already mentioned, this book is primarily devoted to solving equations in Hilbert spaces. The operators involved are usually compact with unbounded inverses. The main applications we have in mind are solutions of integral equations, parameter identification problems and tomographic problems. Other areas, like the solution of ill-posed partial differential equations, like degenerate partial differential equations, are not considered and should be tackled with different methods. There the partial differential operator is in general unbounded and even the evaluation of the forward problem is ill-posed in our sense. We refer to the recent monograph [54] by Groetsch that applies to the stable evaluation of differential operators.

Bibliography

[1] V. Akcelik, G. Biros, A. Draganescu, J. Hill, O. Ghattas, and B. Van Bloemen Waanders, *Dynamic data-driven inversion for terascale simulations: Real-time identification of airborne contaminants*. Proceedings of SC05. IEEE/ACM, Seattle, 2005.

[2] O. M. Alifanov and S. V. Rumyantsev, *On the stability of iterative methods for the solution of linear ill-posed problems*, Soviet Math. Dokl. 20 (1979), pp. 1133–1136.

[3] A. B. Bakushinsky, *The problem of the convergence of the iteratively regularized Gauss-Newton method*, Comput. Math. Math. Phys. 32 (1992), pp. 1353–1359.

[4] A. B. Bakushinsky and A. Goncharsky, *Iterative Methods for the Solution of Incorrect Problems*. Nauka, Moscow, 1989, in Russian.

[5] A. B. Bakushinsky and M. Y. Kokurin, *Iterative Methods for Approximate Solution of Inverse Problems*, Mathematics and Its Applications 577. Springer, Dordrecht, 2004.

[6] R. E. Bank and D. J. Rose, *Analysis of a multilevel iterative method for nonlinear finite element equations*, Math. Comput. 39 (1982), pp. 453–465.

[7] S. P. Banks, *On the solution of Fredholm integral equations of the first kind in L^2*, J. Inst. Math. Appl. 20 (1977), pp. 143–150.

[8] F. Bauer and T. Hohage, *A Lepskij-type stopping rule for regularized Newton methods*, Inverse Problems 21 (2005), pp. 1975–1991.

[9] M. P. Bendsøe and O. Sigmund, *Topology Optimization: Theory, Methods and Applications*. Springer, Berlin, 2003.

[10] A. Binder, M. Hanke, and O. Scherzer, *On the Landweber iteration for nonlinear ill-posed problems*, J. Inv. Ill-Posed Problems 4 (1996), pp. 381–389.

[11] B. Blaschke, A. Neubauer, and O. Scherzer, *On convergence rates for the iteratively regularized Gauss-Newton method*, IMA Journal of Numer. Anal. 17 (1997), pp. 421–436.

[12] L. Borcea, *Electrical impedance tomography*, Inverse Problems 18 (2002), pp. R99–R136.

[13] J. Bramble, *Multigrid methods*, Pitman Research Notes in Mathematics. Longman Scientific & Technical, Harlow, 1993.

[14] A. Brandt, *Multilevel adaptive solutions to boundary value problems*, Math. Comp. 31 (1977), pp. 333–390.

[15] H. Brezis, *Operateurs Maximaux Monotones et semi-groupes de contractions dans les espaces de Hilbert*. North-Holland, Amsterdam, 1973.

[16] F. Browder and W. V. Petryshyn, *Construction of fixed poins of nonlinear mappings in Hilbert space*, J. Math. Anal. Appl. 20 (1967), pp. 197–228.

[17] C. G. Broyden, *A class of methods for solving nonlinear simultaneous equations*, Math. Comp. 19 (1965), pp. 577–593.

[18] G. Bruckner, S. Prößdorf, and G. Vainikko, *Error bounds of discretization methods for boundary integral equations with noisy data*, Appl. Anal. 63 (1996), pp. 25–37.

[19] H. D. Bui and Q. S. Nguyen (eds.), *Thermomechanical Couplings in Solids: IUTAM Symposium, Paris, 1986*. North-Holland, Amsterdam, 1987.

[20] M. Burger, *Levenberg-Marquardt level set methods for inverse obstacle problems*, Inverse Problems 20 (2004), pp. 259–282.

[21] M. Burger, G. Gilboa, S. Osher, and J. Xu, *Nonlinear inverse scale space methods for image restoration*, Comm. Math. Sciences 4 (2006), pp. 179–212.

[22] M. Burger and B. Kaltenbacher, *Regularizing Newton-Kaczmarz methods for nonlinear ill-posed problems*, SIAM J. Numer. Anal. 44 (2006), pp. 153–182.

[23] M. Burger and W. Mühlhuber, *Iterative regularization of parameter identification problems by sequential quadratic programming methods*, Inverse Problems 18 (2002), pp. 943–969.

[24] ———, *Numerical approximation of an SQP-type method for parameter identification*, SIAM J. Numer. Anal. 40 (2002), pp. 1775–1797.

[25] M. Burger and S. Osher, *A survey on level set methods for inverse problems and optimal design*, Eur. J. Appl. Math. 16 (2005), pp. 263–301.

[26] T. Charlebois and R. Pelton, *Quantitative 2D and 3D Schlieren imaging for acoustic power and intensity measurements*, Medical Electronics (1995), pp. 789–792.

[27] G. Chavent, *New size $\times$ curvature conditions for strict quasiconvexity of sets*, SIAM J. Control Optim. 29 (1991), pp. 1348–1372.

[28] ———, *Quasi-convex sets and size $\times$ curvature condition, applications to nonlinear inversion*, Appl. Math. Optim. 24 (1991), pp. 129–169.

[29] G. Chavent and K. Kunisch, *State space regularization: geometric theory*, Appl. Math. Optim. 37 (1998), pp. 243–267.

[30] P. G. Ciarlet, *The Finite Element Method for Elliptic Problems*. North Holland, Amsterdam, 1978.

[31] I. Cioranescu, *Geometry of Banach Spaces, Duality Mappings and Nonlinear Problems*, Mathematics and its Applications 62. Kluwer, Dordrecht, 1990.

[32] F. Colonius and K. Kunisch, *Stability for parameter estimation in two point boundary value problems*, J. Reine Angew. Math. 370 (1986), pp. 1–29.

[33] I. Daubechies, M. Defrise, and C. De Mol, *An iterative thresholding algorithm for linear inverse problems with a sparsity constraint*, Comm. Pure Appl. Math. 57 (2004), pp. 1413–1457.

[34] M. Defrise and C. De Mol, *A note on stopping rules for iterative regularization methods and filtered SVD*. Inverse Problems: An Interdisciplinary Study (P. C. Sabatier, ed.), pp. 261–268. Academic Press, London, Orlando, San Diego, New York, 1987.

[35] M. C. Delfour and J.-P. Zolésio, *Shape analysis via oriented distance functions*, J. Funct. Anal. 123 (1994), pp. 129–201.

[36] J. E. Dennis and J. J. Moré, *Quasi-Newton methods, motivation and theory*, SIAM Review 19 (1977), pp. 46–89.

[37] P. Deuflhard, H. W. Engl, and O. Scherzer, *A convergence analysis of iterative methods for the solution of nonlinear ill-posed problems under affinely invariant conditions*, Inverse Problems 14 (1998), pp. 1081–1106.

[38] P. Deuflhard and G. Heindl, *Affine invariant convergence theorems for Newton's method and extensions to related methods*, SIAM J. Numer. Anal. 16 (1979), pp. 1–10.

[39] P. Deuflhard and F. Potra, *Asymptotic mesh independence of Newton-Galerkin methods via a refined Mysovskii theorem*, SIAM J. Numer. Anal. 29 (1992), pp. 1395–1412.

[40] H. Egger, *Fast fully iterative Newton-type methods for inverse problems*, J. Inv. Ill-Posed Problems 15 (2007), pp. 257–275.

[41] ———, *Y-Scale regularization*, SIAM J. Numer. Anal. 46 (2008), pp. 419–436.

[42] H. Egger and A. Neubauer, *Preconditioning Landweber iteration in Hilbert scales*, Numer. Math. 101 (2005), pp. 643–662.

[43] B. Eicke, *Iteration methods for convexly constrained ill-posed problems in Hilbert space*, Numer. Funct. Anal. Optim. 13 (1992), pp. 413–429.

[44] B. Eicke, A. K. Louis, and R. Plato, *The instability of some gradient methods for ill-posed problems*, Numer. Math. 58 (1990), pp. 129–134.

[45] H. W. Engl, M. Hanke, and A. Neubauer, *Regularization of Inverse Problems*. Kluwer, Dordrecht, 1996.

[46] H. W. Engl and A. Neubauer, *On projection methods for solving linear ill-posed problems*. Model Optimization in Exploration Geophysics (A. Vogel, ed.), pp. 73–92. Vieweg, Braunschweig, 1987.

[47] H. W. Engl and J. Zou, *A new approach to convergence rate analysis of Tiknonov regularization for parameter identification in heat conduction*, Inverse Problems 16 (2000), pp. 1907–1923.

[48] L. C. Evans, *Partial differential equations*, Graduate Studies in Mathematics 19. AMS, Providence, RI, 1998.

[49] L. C. Evans and R. F. Gariepy, *Measure Theory and Fine Properties of Functions*. CRC-Press, Boca Raton, 1992.

[50] S. F. Gilyazov, *Iterative solution methods for inconsistent linear equations with non self-adjoint operators*, Moscow Univ. Comput. Math. Cybernet 78 (1977), pp. 8–13.

[51] P. Gould, *The rise and rise of medical imaging*, Medical Physics (2003), pp. 29–31.

[52] A. Griewank, *Rates of convergence for secant methods on nonlinear problems in Hilbert space*. Numerical Analysis, Proceedings Guanajuato, Mexico, July 1984, number 1230 in Lecture Notes in Mathematics (J. P. Hennart, ed.), pp. 138–157. Springer, Berlin, 1986.

[53] C. W. Groetsch, *The Theory of Tikhonov Regularization for Fredholm Equations of the First Kind*. Pitman, Boston, 1984.

[54] ———, *Stable approximate evaluation of unbounded operators*, Lecture Notes in Mathematics 1894. Springer, Berlin, 2007.

[55] C. W. Groetsch and A. Neubauer, *Convergence of a general projection method for an operator equation of the first kind*, Houston J. Math. 14 (1988), pp. 201–208.

[56] G. Haase, U. Langer, A. Meyer, and S.V. Nepomnyashikh, *Hierarchical extension operators and local multigrid methods in domain decomposition preconditioners*, East-West J. Numer. Math. 2 (1994), pp. 179–193.

[57] E. Haber, *Quasi-Newton methods for large-scale electromagnetic inverse problems*, Inverse Problems 21 (2005), pp. 305–323.

[58] E. Haber and U. Ascher, *A multigrid method for distributed parameter estimation problems*, Inverse Problems 17 (2001), pp. 1847–1864.

[59] W. Hackbusch, *Multi-Grid Methods and Applications*. Springer, Berlin, 1985.

[60] B. Halpern, *Fixed points of nonexpanding maps*, Bull. Amer. Math. Soc. 73 (1967), pp. 957–961.

[61] M. Haltmeier, R. Kowar, A. Leitao, and O. Scherzer, *Kaczmarz methods for regularizing nonlinear ill-posed equations II: Applications*, Inverse Problems and Imaging 1 (2007), pp. 507–523.

[62] M. Haltmeier, A. Leitao, and O. Scherzer, *Kaczmarz methods for regularizing nonlinear ill-posed equations I: convergence analysis*, Inverse Problems and Imaging 1 (2007), pp. 289–298.

[63] M. Hanke, *A regularization Levenberg-Marquardt scheme, with applications to inverse groundwater filtration problems*, Inverse Problems 13 (1997), pp. 79–95.

[64] ———, *Regularizing properties of a truncated Newton-CG algorithm for nonlinear inverse problems*, Numer. Funct. Anal. Optim. 18 (1997), pp. 971–993.

[65] M. Hanke and C. W. Groetsch, *Nonstationary iterated Tikhonov regularization*, J. Optim. Theory Appl. 98 (1998), pp. 37–53.

[66] M. Hanke, F. Hettlich, and O. Scherzer, *The Landweber iteration for an inverse scattering problem*, in Wang et al. [157].

[67] M. Hanke, A. Neubauer, and O. Scherzer, *A convergence analysis of the Landweber iteration for nonlinear ill-posed problems*, Numer. Math. 72 (1995), pp. 21–37.

[68] M. Hanke and C. R. Vogel, *Two-level preconditioners for regularized inverse problems I: Theory*, Numer. Math. 83 (1999), pp. 385–402.

[69] G. T. Herman, *Image Reconstruction from Projections: The Fundamentals of Tomography*. Academic Press, New York, 1980.

[70] A. O. Hero and H. Krim, *Mathematical Methods in Imaging*, IEEE Signal Processing Magazine 19 (2002), pp. 13–14.

[71] F. Hettlich, *An iterative method for the inverse scattering problem from sound-hard obstacles*, ZAMM 76 (1996), pp. 165–168.

[72] ———, *The Landweber iteration applied to inverse conductive scattering problems*, Inverse Problems 14 (1998), pp. 931–947.

[73] F. Hettlich and W. Rundell, *Iterative methods for the reconstruction of an inverse potential problem*, Inverse Problems 12 (1996), pp. 251–266.

[74] ———, *Recovery of the support of a source term in an elliptic differential equation*, Inverse Problems 13 (1997), pp. 959–976.

[75] ———, *The determination of a discontinuity in a conductivity from a single boundary measurement*, Inverse Problems 14 (1998), pp. 67–82.

[76] ———, *A second degree method for nonlinear inverse problems*, SIAM J. Numer. Anal. 37 (2000), pp. 587–620.

[77] T. Hohage, *Logarithmic convergence rates of the iteratively regularized Gauß-Newton method for an inverse potential and an inverse scattering problem*, Inverse Problems 13 (1997), pp. 1279–1299.

[78] ———, *Convergence rates of a regularized Newton method in sound-hard inverse scattering*, SIAM J. Numer. Anal. 36 (1999), pp. 125–142.

[79] ———, *Iterative Methods in Inverse Obstacle Scattering: Regularization Theory of Linear and Nonlinear Exponentially Ill-Posed Problems*, Ph.D. thesis, University of Linz, 1999.

[80] ———, *Regularization of exponentially ill-posed problems*, Numer. Funct. Anal. Optim. 21 (2000), pp. 439–464.

[81] T. Hohage and C. Schormann, *A Newton-type method for a transmission problem in inverse scattering*, Inverse Problems 14 (1998), pp. 1207–1228.

[82] J. Kaipio and E. Somersalo, *Statistical and Computational Inverse Problems*, Applied Mathematical Sciences 160. Springer, New York, 2005.

[83] B. Kaltenbacher, *Some Newton type methods for the regularization of nonlinear ill-posed problems*, Inverse Problems 13 (1997), pp. 729–753.

[84] ———, *On Broyden's method for ill-posed problems*, Numer. Funct. Anal. Optim. 19 (1998), pp. 807–833.

[85] ———, *A posteriori parameter choice strategies for some Newton type methods for the regularization of nonlinear ill-posed problems*, Numer. Math. 79 (1998), pp. 501–528.

[86] ———, *A projection-regularized Newton method for nonlinear ill-posed problems and its application to parameter identification problems with finite element discretization*, SIAM J. Numer. Anal. 37 (2000), pp. 1885–1908.

[87] ———, *Regularization by projection with a posteriori discretization level choice for linear and nonlinear ill-posed problems*, Inverse Problems 16 (2000), pp. 1523–1539.

[88] ———, *On the regularizing properties of a full multigrid method for ill-posed problems*, Inverse Problems 17 (2001), pp. 767–788.

[89] ———, *V-cycle convergence of some multigrid methods for ill-posed problems*, Math. Comp. 72 (2003), pp. 1711–1730.

[90] B. Kaltenbacher, M. Kaltenbacher, and S. Reitzinger, *Identification of nonlinear B-H curves based on magnetic field computations and multigrid methods for ill-posed problems*, Eur. J. Appl. Math. 14 (2003), pp. 15–38.

[91] B. Kaltenbacher and A. Neubauer, *Convergence of projected iterative regularization methods for nonlinear problems with smooth solutions*, Inverse Problems 22 (2006), pp. 1105–1119.

[92] B. Kaltenbacher, A. Neubauer, and A. G. Ramm, *Convergence rates of the continuous regularized Gauss-Newton method*, J. Inv. Ill-Posed Problems 10 (2002), pp. 261–280.

[93] B. Kaltenbacher and J. Schicho, *A multi-grid method with a priori and a posteriori level choice for the regularization of nonlinear ill-posed problems*, Numer. Math. 93 (2002), pp. 77–107.

[94] J. T. King, *Multilevel algorithms for ill-posed problems*, Numer. Math. 61 (1992), pp. 311–334.

[95] A. Kirsch, *The domain derivative and two applications in inverse scattering theory*, Inverse Problems 9 (1993), pp. 81–96.

[96] ———, *An Introduction to the Mathematical Theory of Inverse Problems*. Springer, New York, 1996.

[97] R. Kohn and G. Strang, *Optimal design and relaxation of variational problems. I*, Comm. Pure Appl. Math. 39 (1986), pp. 113–137.

[98] ———, *Optimal design and relaxation of variational problems. II*, Comm. Pure Appl. Math. 39 (1986), pp. 139–182.

[99] ———, *Optimal design and relaxation of variational problems. III*, Comm. Pure Appl. Math. 39 (1986), pp. 353–377.

[100] K. Kowalenko, *Saving lives, one land mine at a time*, IEEE, The Institute 28 (2004), pp. 1,10–11.

[101] R. Kowar and O. Scherzer, *Convergence analysis of a Landweber-Kaczmarz method for solving nonlinear ill-posed problems*. Ill-Posed and Inverse Problems (V. G. Romanov, S. I. Kabanikhin, Yu. E. Anikonov, and A. L. Bukhgeim, eds.), pp. 69–90. VSP, Zeist, 2002.

[102] S. G. Krein and J. I. Petunin, *Scales of Banach spaces*, Russian Math. Surveys 21 (1966), pp. 85–160.

[103] R. Kress and W. Rundell, *A quasi-Newton method in inverse obstacle scattering*, Inverse Problems 10 (1994), pp. 1145–1157.

[104] R. A. Kruger, W. L. Kiser, K. D. Miller, and H. E. Reynolds, *Thermoacoustic CT: imaging principles*, Proc SPIE 3916 (2000), pp. 150–159.

[105] R. A. Kruger, D. R. Reinecke, and G. A. Kruger, *Thermoacoustic computed tomography – technical considerations*, Medical Physics 26 (1999), pp. 1832–1837.

[106] R. A. Kruger, K. M. Stantz, and W. L. Kiser, *Thermoacoustic CT of the Breast*, Proc. SPIE 4682 (2002), pp. 521–525.

[107] P. Kügler, *A derivative free Landweber iteration for parameter identification in certain elliptic PDEs*, Inverse Problems 19 (2003), pp. 1407–1426.

[108] ———, *A Derivative Free Landweber Method for Parameter Identification in Elliptic Partial Differential Equations with Application to the Manufacture of Car Windshields*, Ph.D. thesis, Johannes Kepler University, Linz, Austria, 2003.

[109] ———, *Identification of a temperature dependent heat conductivity from single boundary measurements*, SIAM J. Numer. Anal. 41 (2003), pp. 1543–1563.

[110] L. Landweber, *An iteration formula for Fredholm integral equations of the first kind*, Amer. J. Math. 73 (1951), pp. 615–624.

[111] V. G. Maz'ya, *Sobolev Spaces*. Springer, Berlin, Heidelberg, New York, 1985.

[112] S. F. McCormick, *An iterative procedure for the solution of constrained nonlinear equations with applications to optimization problems*, Numer. Math. 23 (1975), pp. 371–385.

[113] ———, *The methods of Kaczmarz and row orthogonalization for solving linear equations and least squares problems in Hilbert space*, Indiana Univ. Math. J. 26 (1977), pp. 1137–1150.

[114] S. F. McCormick and G. H. Rodrigue, *A uniform approach to gradient methods for linear operator equations*, J. Math. Anal. Appl. 49 (1975), pp. 275–285.

[115] K. H. Meyn, *Solution of underdetermined nonlinear equations by stationary iteration methods*, Numer. Math. 42 (1983), pp. 161–172.

[116] V. A. Morozov, *On the solution of functional equations by the method of regularization*, Soviet Math. Dokl. 7 (1966), pp. 414–417.

[117] M. Z. Nashed (ed.), *Generalized Inverses and Applications*. Academic Press, New York, 1976.

[118] M. Z. Nashed and X. Chen, *Convergence of Newton-like methods for singular operator equations using outer inverses*, Numer. Math. 66 (1993), pp. 235–257.

[119] F. Natterer, *Regularisierung schlecht gestellter Probleme durch Projektionsverfahren*, Numer. Math. 28 (1977), pp. 329–341.

[120] ———, *Error bounds for Tikhonov regularization in Hilbert scales*, Appl. Anal. 18 (1984), pp. 29–37.

[121] ———, *The Mathematics of Computerized Tomography*. SIAM, Philadelphia, 2001.

[122] F. Natterer and F. Wübbeling, *Mathematical Methods in Image Reconstruction*. Society for Industrial and Applied Mathematics (SIAM), Philadelphia, PA, 2001.

[123] A. Neubauer, *Tikhonov regularization of nonlinear ill-posed problems in Hilbert scales*, Appl. Anal. 46 (1992), pp. 59–72.

[124] ———, *On Landweber iteration for nonlinear ill-posed problems in Hilbert scales*, Numer. Math. 85 (2000), pp. 309–328.

[125] A. Neubauer and O. Scherzer, *A convergent rate result for a steepest descent method and a minimal error method for the solution of nonlinear ill-posed problems*, ZAA 14 (1995), pp. 369–377.

[126] S. Osher and R. Fedkiw, *Level set methods and dynamic implicit surfaces*, Applied Mathematical Sciences 153. Springer, New York, 2003.

[127] S. Osher and J. A. Sethian, *Fronts propagating with curvature-dependent speed: Algorithms based on Hamilton-Jacobi formulations*, J. Comput. Phys. 79 (1988), pp. 12–49.

[128] C. Pechstein and B. Jüttler, *Monotonicity-preserving interproximation for BH curves*, J. Comp. Appl. Math. 196 (2006), pp. 45–57.

[129] S. V. Pereverzev and S. Prößdorf, *On the characterization of self-regularization properties of a fully discrete projection method for Symm's integral equation*, J. Integral Equations Appl. 12 (2000), pp. 113–130.

[130] W. V. Petryshyn and T. E. Williamson, *Strong and weak convergence of the sequence of successive approximations for quasi-nonexpansive mappings*, J. Math. Anal. Appl. 43 (1973), pp. 459–497.

[131] T. A. Pitts, J. F. Greenleaf, Jian yu Lu, and R. R. Kinnick, *Tomographic Schlieren imaging for measurment of beam pressure and intensity*, Proc. IEEE Ultrasonics Symposium (1994), pp. 1665–1668.

[132] R. Plato and G. Vainikko, *On the regularization of projection methods for solving ill-posed problems*, Numer. Math. 57 (1990), pp. 63–79.

[133] J. Radon, *Über die Bestimmung von Funktionen durch ihre Integralwerte längs gewisser Mannigfaltigkeiten* (reprinted). J. Radon, Gesammelte Abhandlungen – Collected Works, Band 2 (P. M. Gruber, E. Hlawka, W. Nöbauer, and L. Schmetterer, eds.). Verlag d. Österr. Akad. d. Wiss., Birkhäuser, Basel, 1987.

[134] R. Ramlau, *A modified Landweber-method for inverse problems*, Numer. Funct. Anal. Optim. 20 (1999), pp. 79–98.

[135] A. G. Ramm and A. B. Smirnova, *A numerical method for solving nonlinear ill-posed problems*, Numer. Funct. Anal. Optim. 20 (1999), pp. 317–332.

[136] A. Rieder, *A wavelet multilevel method for ill-posed problems stabilized by Tikhonov regularization*, Numer. Math. 75 (1997), pp. 501–522.

[137] ______, *On convergence rates of inexact Newton regularizations*, Numer. Math. 88 (2001), pp. 347–365.

[138] ______, *Inexact Newton regularization using conjugate gradients as inner iteration*, SIAM J. Numer. Anal. 43 (2005), pp. 604–622.

[139] E. W. Sachs, *Broyden's method in Hilbert space*, Mathematical Programming 35 (1986), pp. 71–82.

[140] F. Santosa, *A level-set approach for inverse problems involving obstacles*, ESAIM Contrôle Optim. Calc. Var. 1 (1995/96), pp. 17–33.

[141] O. Scherzer, *Convergence criteria of iterative methods based on Landweber iteration for nonlinear problems*, J. Math. Anal. Appl. 194 (1995), pp. 911–933.

[142] ______, *A convergence analysis of a method of steepest descent and a two-step algorithm for nonlinear ill-posed problems*, Numer. Funct. Anal. Optim. 17 (1996), pp. 197–214.

[143] ______, *An iterative multi level algorithm for solving nonlinear ill-posed problems*, Numer. Math. 80 (1998), pp. 570–600.

[144] ______, *A modified Landweber iteration for solving parameter estimation problems*, Appl. Math. Optim. 38 (1998), pp. 45–68.

[145] O. Scherzer, M. Grasmair, H. Grossauer, M. Haltmeier, and F. Lenzen, *Variational Methods in Imaging*. Springer, New York, 2008, under final review.

[146] F. Schöpfer, A. K. Louis, and T. Schuster, *Nonlinear iterative methods for linear ill-posed problems in Banach spaces*, Inverse Problems 22 (2006), pp. 311–329.

[147] T. I. Seidman, *Nonconvergence results for the application of least-squares estimation to ill-posed problems*, J. Optim. Theory Appl. 30 (1980), pp. 535–547.

[148] J. A. Sethian, *Level set methods and fast marching methods*, 2nd. ed.. Cambridge University Press, Cambridge, 1999.

[149] M. Tanaka and H. D. Bui (eds.), *Inverse Problems in Engineering Mechanics: IUTAM Symposium, Tokyo, 1992*. Springer, Berlin, 1993.

[150] U. Tautenhahn, *On the asymptotical regularization of nonlinear ill-posed problems*, Inverse Problems 10 (1994), pp. 1405–1418.

[151] G. Vainikko and U. Hämarik, *Projection methods and self-regularization in ill-posed problems*, Sov. Math. 29 (1985), pp. 1–20.

[152] G. M. Vainikko, *Error estimates of the successive approximation method for ill-posed problems*, Automat. Remote Control 40 (1980), pp. 356–363.

[153] G. M. Vainikko and A. Y. Veretennikov, *Iteration Procedures in Ill-Posed Problems*. Nauka, Moscow, 1986, in Russian.

[154] V. V. Vasin, *Iterative methods for solving ill-posed problems with a priori information in Hilbert spaces*, USSR Comput. Math. Math. Phys. 28 (1988), pp. 6–13.

[155] C. R. Vogel, *Computational Methods for Inverse Problems*. SIAM, Providence, 2002.

[156] E. Vonderheid, *Seeing the invisible*, IEEE, The Institute 27 (2003), pp. 1,13.

[157] K.-W. Wang et al. (eds.), *Proceedings of the 1995 Design Engineering Technical Conferences, Vibration Control, Analysis, and Identification, Vol. 3, Part C*. The American Society of Mechanical Engineers, New York, 1995.

[158] S. Webb (ed.), *The Physics of Medical Imaging*. Institute of Physics Publishing, Bristol, Philadelphia, 2000, reprint of the 1988 edition.

[159] R. West, *In industry, seeing is believing*, Physics World (2003), pp. 27–30.

[160] M. Xu and L.-H. V. Wang, *Analytic explanation of spatial resolution related to bandwidth and detector aperture size in thermoacoustic or photoacoustic reconstruction*, Physical Review E 67 (2003), pp. 1–15.

[161] E. Zeidler, *Nonlinear Functional Analysis and its Applications II/B*. Springer, New York, 1990.

Index

R

S

T